Yasin Elshorbany

Tropospheric Oxidation Capacity and Ozone Photochemical Formation

Yasin Elshorbany

Tropospheric Oxidation Capacity and Ozone Photochemical Formation

Investigation of the current understanding of urban atmospheric chemistry: Field Measurements and Modeling Study in the city of Santiago de Chile

Südwestdeutscher Verlag für Hochschulschriften

Impressum/Imprint (nur für Deutschland/ only for Germany)
Bibliografische Information der Deutschen Nationalbibliothek: Die Deutsche Nationalbibliothek verzeichnet diese Publikation in der Deutschen Nationalbibliografie; detaillierte bibliografische Daten sind im Internet über http://dnb.d-nb.de abrufbar.

Alle in diesem Buch genannten Marken und Produktnamen unterliegen warenzeichen-, marken- oder patentrechtlichem Schutz bzw. sind Warenzeichen oder eingetragene Warenzeichen der jeweiligen Inhaber. Die Wiedergabe von Marken, Produktnamen, Gebrauchsnamen, Handelsnamen, Warenbezeichnungen u.s.w. in diesem Werk berechtigt auch ohne besondere Kennzeichnung nicht zu der Annahme, dass solche Namen im Sinne der Warenzeichen- und Markenschutzgesetzgebung als frei zu betrachten wären und daher von jedermann benutzt werden dürften.

Verlag: Südwestdeutscher Verlag für Hochschulschriften Aktiengesellschaft & Co. KG
Dudweiler Landstr. 99, 66123 Saarbrücken, Deutschland
Telefon +49 681 37 20 271-1, Telefax +49 681 37 20 271-0
Email: info@svh-verlag.de
Zugl.: Wuppertal Universität, 2010

Herstellung in Deutschland:
Schaltungsdienst Lange o.H.G., Berlin
Books on Demand GmbH, Norderstedt
Reha GmbH, Saarbrücken
Amazon Distribution GmbH, Leipzig
ISBN: 978-3-8381-1638-9

Imprint (only for USA, GB)
Bibliographic information published by the Deutsche Nationalbibliothek: The Deutsche Nationalbibliothek lists this publication in the Deutsche Nationalbibliografie; detailed bibliographic data are available in the Internet at http://dnb.d-nb.de.

Any brand names and product names mentioned in this book are subject to trademark, brand or patent protection and are trademarks or registered trademarks of their respective holders. The use of brand names, product names, common names, trade names, product descriptions etc. even without a particular marking in this works is in no way to be construed to mean that such names may be regarded as unrestricted in respect of trademark and brand protection legislation and could thus be used by anyone.

Publisher: Südwestdeutscher Verlag für Hochschulschriften Aktiengesellschaft & Co. KG
Dudweiler Landstr. 99, 66123 Saarbrücken, Germany
Phone +49 681 37 20 271-1, Fax +49 681 37 20 271-0
Email: info@svh-verlag.de

Printed in the U.S.A.
Printed in the U.K. by (see last page)
ISBN: 978-3-8381-1638-9

Copyright © 2010 by the author and Südwestdeutscher Verlag für Hochschulschriften Aktiengesellschaft & Co. KG and licensors
All rights reserved. Saarbrücken 2010

To My Parents and the spirit of my Brother Ahmad

and

To My Wife and my lovely Daughter Lina

Acknowledgment

First, I would like to express my sincere gratitude to Prof. Dr. Peter Wiesen for his scientific supervision, continuous support and encouragement. I would like to thank him also for giving me the chance to do my research work in his group and to participate in the different workshops and conferences in Europe and the USA.

I am also grateful to PD Dr. Jörg Kleffmann for introducing me to the HONO measurements by the LOPAP instrument, for the constructive interactive discussion during our different laboratory experiments, field campaigns and data evaluations. I would like to thank him also for co-refereeing this thesis.

My thanks extend to Dr. Ralf Kurtenbach for introducing me to the GC analysis techniques and for the active contribution during our different field campaigns.

I am deeply indebted to Prof. Michael Pilling and Dr. Andrew Rickard from the University of Leeds for introducing me to the Master Chemical Mechanism, MCM, for the very helpful discussion during the preparation and publication of my modelling results.

I would like also to thanks all the participants in the measurements campaigns in Santiago de Chile. My thanks extend also to my colleagues, friends, technical stuff and secretary at the Physical Chemisrty Laboratory of the University of Wuppertal. I would like also to express my sincere gratitude to my colleagues and friends at the National Research Centre in Egypt.

Last and not least, I would like to express my sincere gratitude to my great parents, for their encouragements and support during all the stages of my life. I would like also to express my sincere gratitude to my wife for her patience, support and encouragement during the preparation of this work and for my lovely daughter Lina for her lovely smile, which gave me a lot of support to work further.

Abstract

The oxidation capacity and ozone photochemical formation of the highly polluted urban area of Santiago de Chile has been evaluated during two field measurements campaigns during summer and winter from March 8 – 20 and from May 25 – June 07, 2005, respectively. The OH radical budget was evaluated in both campaigns employing a simple quasi-photo stationary-state model (PSS) constrained with simultaneous measurements of HONO, HCHO, O_3, PAN, NO, NO_2, $j(O^1D)$, $j(NO_2)$, 13 alkenes and meteorological parameters. In addition, a zero dimensional photochemical box model based on the Master Chemical Mechanism (MCMv3.1) has been used for the analysis of the radical budgets and concentrations of OH, HO_2 and RO_2. Besides the above parameters, the MCM model has been constrained by the measured CO and other volatile organic compounds (VOCs) including alkanes and aromatics.

Total production and destruction rates of OH and HO_2 in winter were about two times lower than that during summer. Simulated OH levels by both PSS and MCM models were similar during the daytime for both, summer and winter indicating that the primary OH sources and sinks included in the simple PSS model are predominant. On a 24 h basis, HONO photolysis was shown to be the most important primary OH radical source comprising 52 % and 81 % of the OH initiation rate during summer and winter, respectively followed by alkene ozonolysis (29 % and 12.5 %), photolysis of HCHO (15 % and 6.1 %), and photolysis of O_3 (4 % and <1 %), respectively.

During both summer and winter, there was a balance between the secondary production (HO_2 + NO) and destruction (OH + VOC) of OH radicals indicating that initiation sources of RO_2 and HO_2 are no net OH initiation sources. This result was found to be fulfilled also for other studies investigated. Seasonal impacts on the radical budgets are also discussed.

The photochemical formation of ozone during the summer campaign carried out from March 8 – 20, 2005 has been investigated using an urban photochemical box model based on the Master Chemical Mechanism (MCMv3.1). The model has been constrained with the same set of the measured parameters used to simulate the radical budgets (see above) except O_3, NO_2 and PAN. The O_3-NO_x-VOC sensitivities have been determined by simulating ozone formation at different VOC and NO_x concentrations.

Ozone sensitivity analyses showed that photochemical ozone formation is VOC-limited under average summertime conditions in Santiago. The results of the model simulations have been compared with a set of potential empirical indicator relationships including H_2O_2/HNO_3, $HCHO/NO_y$ and O_3/NO_z. The ozone forming potential of each measured VOC has been determined using the MCM box model. The impacts of the above study on possible summertime ozone control strategies in Santiago are discussed.

Contents

1 Introduction 1
    1.1   **HO_x Tropospheric Chemistry** 4
        1.1.1   Radical Initiation 4
        1.1.2   Radical Propagation 9
        1.1.3   Radical Termination 10
        1.1.4   OH reactivity 11
    1.2   **Tropospheric Modelling Techniques** 12
    1.3   **Oxidation Capacity** 13
    1.4   **Photochemical Ozone Sensitivity** 17
    1.5   **Atmospheric Chemistry Studies in Santiago de Chile** 22
    1.6   **Objectives** 27

2 Methodology 29
    2.1   **Measurements Site** 29
    2.2   **Measurement Techniques** 30
        2.2.1   HONO Measurements 31
            2.2.1.1   Analytical Method 31
            2.2.1.2   Instrument Calibration 32
            2.2.1.3   Instrument Parameters 33
        2.2.2   HCHO Measurements 33
            2.2.2.1   Analytical Method 34
            2.2.2.2   Instrument Calibration 34
            2.2.2.3   Instrument Parameters 36
        2.2.3   VOC measurements and analysis 36
            2.2.3.1   Samples Collection 37
            2.2.3.2   Offline GC-FID Analysis 37
            2.2.3.3   Calibration Procedure 38
            2.2.3.4   Performance Criteria 40
            2.2.3.5   Calculation of Ambient Concentrations 41
            2.2.3.6   Ozone Interferences 44
        2.2.4   Other Parameters 45
            2.2.4.1   PAN Measurements 45
            2.2.4.2 NO_x Measurements 45

- 2.2.4.3 Photolysis Frequencies ... 45
- 2.2.4.4 CO Measurements ... 46
- 2.2.4.5 CO_2 Measurements ... 46
- 2.2.4.6 Ozone Measurements ... 46
- 2.2.4.7 Meteorological Parameters ... 46
- **2.3 Modelling Approach** ... 47
  - 2.3.1 Simple Quasi-Photo-Stationary State Model, PSS ... 47
  - 2.3.2 The Master Chemical Mechanism, MCM ... 48
    - 2.3.2.1 Simulation of Radical budgets ... 48
    - 2.3.2.2 Photochemical Simulation of Ozone ... 49

3 Results and Discussion ... 53
- **3.1 Measurements Results Analysis** ... 53
  - 3.1.1 Summer Campaign ... 53
  - 3.1.2 Winter Campaign and Comparison to Summer ... 56
  - 3.1.3 Emission Indices and Emission Ratios ... 62
  - 3.1.4 Conclusion ... 69
- **3.2 Oxidation Capacity and its Seasonal Dependence** ... 71
  - 3.2.1 Oxidation Capacity ... 71
  - 3.2.2 Radical Production and Destruction Rates ... 72
  - 3.2.3 OH Reactivity ... 78
  - 3.2.4 Radical Propagation ... 79
  - 3.2.5 Balance Ratio and Comparison with other Studies ... 83
  - 3.2.6 Net Radical Sources ... 87
  - 3.2.7 Simulated OH Levels ... 92
  - 3.2.8 Correlation of OH and P_R with $j(O^1D)$ and $j(NO_2)$ ... 94
  - 3.2.9 Source Apportionments of the main OH Radical Precursors ... 95
    - 3.2.9.1 Formaldehyde (HCHO) Contribution ... 95
    - 3.2.9.2 Alkene Ozonolysis Contribution ... 100
    - 3.2.9.3 HONO Daytime Contribution ... 102
    - 3.2.9.4 HONO Dark Sources: ... 104
  - 3.2.10 Conclusion ... 105
- **3.3 Summertime Photochemical Ozone Formation in Santiago, Chile** ... 107
  - 3.3.1 Simulated Ozone Levels and Model Limitations ... 107
  - 3.3.2 VOC/NO_x Ratio ... 110

	3.3.3	Ozone Sensitivity Analysis using the MCM Box Model 110
	3.3.4	Empirical Indicator Species/Relationships 113
	3.3.5	Ozone Production Efficiency, OPE 114
	3.3.6	Photochemical Ozone Production 115
	3.3.7	Photochemical Incremental Reactivity (PIR) Scale 119
	3.3.8	Ozone Diurnal Structure 121
	3.3.9	Nitrous Acid Contribution to Ozone Formation 124
	3.3.10	Impacts on the Ozone Control Strategy 125
	3.3.11	Conclusion 127

4   Summary 129

5   Outlook 133

   5.1   Radical Budgets: 133

   5.2   Radical Balance: 134

   5.3   Ozone Control Strategy: 134

6   Appendices 137

   6.1   Appendix A: Measurement site 137

   6.2   Appendix B: VOC sampling system and calibration data 139

   6.3   Appendix C: VOC measurements 144

7   Bibliography 149

1   Introduction

The atmosphere is a dynamic system, with its gaseous constituents continuously being exchanged with vegetation, ground surfaces, the oceans, and biological organisms. Chemical processes within the atmosphere itself, by biological activity, volcanic exhalation, radioactive decay, and human industrial activities, produce gases. Gases are removed from the atmosphere by chemical reactions in the atmosphere, by biological activity, by physical processes (such as particle formation) in the atmosphere, and by deposition and uptake by the oceans and earth. The average lifetime of a gas molecule introduced into the atmosphere can range from seconds to many years, depending on the effectiveness of the removal processes (Seinfeld and Pandis, 1998). The Earth's atmosphere is composed primarily of N_2 and O_2 and several noble gases. The remaining gaseous constituents, the trace gases, comprise less than 1 % of the atmosphere. These trace gases play a crucial role in the earth's radiative balance and in the chemical properties of the atmosphere (Seinfeld and Pandis, 1998).

Atmospheric Chemistry is an interdisciplinary science, which incorporates diverse areas of chemistry including photochemistry, spectroscopy, kinetics and mechanism of homogeneous and heterogeneous organic and inorganic reactions (Finlayson Pitts and Pitts, 2000). Atmospheric chemistry as a scientific discipline goes back to the 18th century when the principle issue was identifying the major chemical components of the atmosphere, nitrogen, oxygen, water, carbon monoxide and noble gases (Seinfeld and Pandis, 1998). Atmospheric chemistry range from field measurements of trace gas species, laboratory studies, and different techniques of photochemical modelling to the health and environmental risk assessment and associated risk management decisions for the control of air pollutants.

Atmospheric chemistry focuses on the investigation of the atmospheric chemical composition in the region closest to the earth's surface, the troposphere ($\leq$10-15 km), but also to the tropopause (~10) and the stratosphere (~10-50 km) (Finlayson Pitts and Pitts, 2000). The troposphere is the lowest layer of the atmosphere and can be seen as two different layers, the lower layer, which is in close contact with the Earth's surface, is the planetary boundary layer (PBL) while the one above is the free troposphere (FT). The troposphere is also characterized by a negative temperature gradient that results in rapid vertical mixing of gases. The vertical motion within the atmosphere results from 1) convection from solar heating of the Earth's surface, 2) convergence or divergence of horizontal flows, 3) horizontal flow over topographic features at the earth's surface and 4) buoyancy caused by the release of latent heat as water condenses (Seinfeld and Pandis, 1998).

It has become increasingly clear that the troposphere and stratosphere are intimately connected. This is demonstrated by the vertical transport of long-lived ozone-destroying anthropogenic emissions of chlorofluorocarbons (CFCs) and conversely the downward transport of stratospheric ozone into the troposphere.

The Chapman cycle hypothesized in the 1930 by Sir Sydney Chapman is responsible for generating steady-state concentrations of O_3 in the stratosphere.

R 1.1 O_2 + $h\nu$ → $2O$

R 1.2 O + O_2 → O_3

R 1.3 O + O_3 → $2O_2$

R 1.4 O_3 + $h\nu$ → O + O_2

Stratospheric ozone is essential for the life on the earth because it strongly absorbs light of λ <290 nm. As a result, sunlight reaching the troposphere, commonly referred to as actinic radiation, has wavelengths longer than 290 nm. Only those molecules that absorb radiation at wavelengths longer than 290 nm can undergo photo dissociation and other primary photochemical processes.

In spite the fact that the atmosphere is composed predominantly of relatively inert molecules such as N_2 and O_2, it is actually a rather efficient oxidizing medium. The atmosphere's oxidizing capacity is due to the presence of minute amounts of very reactive oxidizing molecular fragments, called free radicals (Seinfeld and Pandis, 1998) in addition to other less reactive oxidizing molecular species.

Introduction

The main oxidizing species in the atmosphere are hydroxyl radicals (OH), nitrate radicals (NO_3) and ozone (O_3) molecules. The hydroxyl radical is the principal oxidizing agent in the atmosphere. It controls and determines the oxidizing power of the atmosphere and thus governs the atmospheric lifetime of many species, and hence their potential to contribute to climate change and ozone depletion. The nitrate radical formed from the oxidation of NO_2 by O_3 photolyses on a time scale of 5 s to return to NO_2. NO_3 is therefore important during the night and can provide an important sink of some unsaturated hydrocarbons (Wayne, 1991). In addition, ozone (O_3) is an important oxidizer of unsaturated hydrocarbons, which also participates in the formation of the hydroxyl radical.

Large emissions of both anthropogenic and biogenic volatile organic compounds are being released to the atmosphere. These emissions may be subjected to local and regional transportation and/or subsequent oxidation processes that may results in the formation of other harmful secondary products such as ozone and PAN, which are the main constituents of the summer photochemical smog in many polluted areas around the world.

Improving our understanding of the abundance and distribution of radicals is thus a key goal of atmospheric chemistry. Among the different radical species, the OH radical plays the dominant role in the degradation of trace gases in the atmosphere. The major role of OH in the atmosphere was first recognised by Levy (1971). Because of its short life time (<1s in the mid-latitude continental boundary layer), OH concentrations are determined by local chemical processes rather than transport. Using an indirect method, the global diurnal mean concentration of OH of $1.1\pm0.2\times10^6$ radicals/cm^3 has been derived using the long-term measurements of methyl chloroform, a compound that is mainly removed from the atmosphere by OH and has only anthropogenic sources (Lelieveld et al., 2004 and references therein). Different methods are being used for tropospheric OH radical measurements. Among which, three methods have enjoyed considerable success. The first two are both spectroscopic methods, namely laser-induced fluorescence spectroscopy at low pressure (LIF-FAGE) and long-path differential optical absorption spectroscopy (DOAS). The third is the chemical ionization mass spectroscopy (CIMS) method, whereby OH is converted chemically into $H_2^{34}SO_4$, a species that does not occur naturally (and hence has no background), and which is subsequently measured by mass spectrometry (Heard and Pilling, 2003). For the detection of

HO_2, three methods have been demonstrated in the field, namely chemical conversion to OH followed by detection of the OH generated using laser-induced fluorescence at low pressure, matrix isolation electron spin resonance (MIESR), and a sulphur-chemistry peroxy-radical amplifier combined with detection via CIMS, which is able to discriminate HO_2 from total peroxy radicals (Heard and Pilling, 2003).

Comparison of in situ measurements of HO_x (OH + HO_2) with model simulations constrained by observations of longer-lived species such as NO_x (NO+NO_2), ozone and VOCs have been used frequently to test our understanding of atmospheric processes, and to validate the chemical mechanisms included in tropospheric models. Understanding the processes and rates by which species are oxidized in the atmosphere, i.e., the oxidizing power of the atmosphere is crucial to our knowledge of atmospheric composition. Changes in the oxidizing power of the atmosphere would have a wide range of implications for air pollution, aerosol formation, greenhouse radiative forcing, and stratospheric ozone depletion (Thomson, 1992).

1.1 HO_x Tropospheric Chemistry

The current understanding of the HO_x (OH + HO_2) tropospheric chemistry can be summarized in terms of three main processes, namely radical initiation, propagation and termination. Radical initiation defines procedures that may lead to production of new radicals. These new radicals initiate the CO and hydrocarbons oxidation process, which lead to production and destruction of radicals and is so called "propagation process". Radical termination includes procedures that may lead to a permanent radical destruction.

1.1.1   Radical Initiation

A major OH production pathway in the troposphere is the photolysis of ozone by a narrow band of radiation in the 290-330 nm range, followed by reaction of the resulting electronically excited O^1D atoms with water vapour, in competition with their collisional quenching. Only ~10 % of the $O(^1D)$ atoms react with H_2O under tropospheric conditions of 1 atm, 50 % relative humidity and 298 K while the most are deactivated to the ground state $O(^3P)$ by collision with a third molecule (i.e., M = N_2 or O_2). Ground state $O(^3P)$ that can also be directly formed from photo-dissociation of O_3 recombines immediately with molecular oxygen to form again O_3. Radiation in this wavelength range is strongly absorbed by overhead O_3 column and hence the production of O^1D is strongly dependent on the thickness of the stratospheric O_3 layer in addition to the solar zenith angle (SZA) (Madronich and Garnier, 1992):

R 1.5 O_3 + hv ($\lambda \leq 340$ nm) → O_2 + $O(^1D)$

R 1.6 $O(^1D)$ + H_2O → 2 OH

R 1.7 $O(^1D)$ + M → $O(^3P)$ + M

R 1.8 $O(^3P)$ + O_2 + M → O_3 + M

HONO photolysis is another major OH source, especially during the early morning (Perner and Platt, 1979):

R 1.9 HONO + hv ($\lambda < 400$nm) → OH + NO

Also during the daytime, photolysis of HONO has recently been shown to be the most important OH initiation source (Kleffmann et al., 2005, Ren et al., 2006; Dusanter et al, 2009). Also in other field work studies, unexpected high daytime values of HONO were observed (e.g. Neftel et al., 1996, Zhou et al., 2002; Kleffmann et al., 2002; Acker et al., 2006a, 2006b) and new photochemical HONO sources have been proposed (Kleffmann, 2007), some of which have recently been identified in the laboratory (e.g., Zhou et al., 2003; George et al., 2005; Stemmler et al., 2006, 2007; Bejan et al., 2006; Li et al., 2008, Gustafsson et al., 2006; Ndour et al., 2008). HONO sources in the atmosphere are not yet well understood. While the dark heterogeneous conversion of NO_2 on humid surfaces (Finlayson-Pitts et al., 2003) is commonly accepted as the dominant HONO sources during night (Alicke et al., 2002), the exact mechanism is still unclear:

R 1.10 $2 NO_2 + H_2O$ → $HONO + HNO_3$

In addition, recent studies showed that the heterogeneous reaction of NO_2 with adsorbed hydrocarbons (R 1.11) is also important under atmospheric conditions (Ammann, et al., 2005).

R 1.11 $NO_2 + Organics_{(red)}$ → $HONO + Organics_{(oxi.)}$

For HONO daytime sources, five photochemical mechanisms were recently identified. Three of them dominate under high NO_x urban conditions, namely heterogeneous conversion of gaseous NO_2 on photosensitized solid surface organic compounds (George et al., 2005;

Stemmler et al., 2006), photocatalytic conversion of NO_2 on TiO_2 (Gustafsson et al., 2006; Ndour et al., 2008) and the photolysis of the gaseous nitroaromatic compounds (Bejan et al., 2006) and should correlate well with $j(NO_2)$. Under low NO_x rural conditions, the photolysis of nitric acid (Zhou et al., 2003) adsorbed on solid surfaces (including vegetations) may dominate and would better correlate to $j(O^1D)$, caused by the much lower wavelength range of the nitric acid photolysis.

A recent study by Li et al. (2008) suggested an additional source of HONO and OH from the reaction of electronically excited nitrogen dioxide (denoted as NO_2^*) by visible light (400 nm < λ > 650 nm) with water vapour:

R 1.12 NO_2 +   hv (λ>420 nm)   → NO_2^*
R 1.13 NO_2^* + H_2O →   HONO + OH
R 1.14 NO_2^* + H_2O → $NO_2 + H_2O$
R 1.15 NO_2^* +   M (N_2 or O_2)   → NO_2 + M

Owing to the very large solar flux in the visible region of the spectrum, Li et al. (2008) found that despite the low OH yield (~0.001) from the reaction NO_2^* with H_2O, reaction R 1.13 represents a major source of OH of up to 50% of that produced from the $j(O^1D)$ reaction with water vapour (reaction R 1.6) at high solar zenith angles under polluted conditions. In the same issue of the science magazine where Li et al. (2008) published their work, Wennberg and Dabdub (2008) published results of model simulation of the air quality in Los Angeles basin for typical summer smog episode. Their model simulations with this new mechanism showed that the ozone concentrations were higher by 30 – 40 % than that obtained without the new mechanism. They showed however, that this new mechanism overestimate the observed O_3 distribution in Los Angeles, and therefore concluded that it is possible that the rate constant by Li et al., (2008) is too large (Wennberg and Dabdub, 2008). It should be also mentioned that a previous study by Crowley and Carl (1997) showed that the OH formation rate from reaction R 1.13 is more than one order of magnitude lower (1.2×10^{-14} cm^3 $molecule^{-1}$ s^{-1}) than that of (1.7×10^{-13} cm^3 $molecule^{-1}$ s^{-1}) reported by Li et al (2008). Crowley and Carl (1997) found also that this process will be inefficient in the troposphere due to the low photolysis intensities and rapid collisional quenching of NO_2^* (R 1.14 - R 1.15) and placed an upper limit of ~2 % to the formation of OH via reaction R 1.13 in the troposphere when solar zenith angles are high. In their comment on the Li et al. (2008) paper, Carr et al. (2009) suggested also that the reaction NO_2^* with H_2O has an OH yield of <0.00006 (17 times lower than that of Li et al., 2008) and thus has little impact on atmospheric chemistry. Recently, Dusanter et al. (2009) included the above mechanism in a RACM-based zero dimensional box model using the rate constant of Li et al. (2008). They found that the relative importance of the reaction of NO_2^* with H_2O to the OH initiation rate is ~1 % and thus is negligible for the urban area of Mexico City where other radical sources dominate (Dusanter et al., 2009). Very recently, Sarwar et al. (2009) examined the impact of the photo-excited NO_2 chemistry on regional air quality in the US and found that excited NO_2 chemistry can increase the monthly mean daytime hydroxyl radical and nitrous acid by a maximum of 28% and 100 pptv, respectively.

In conclusion, the issue of the excited NO_2 chemistry still need further experimental verification concerning the applied rate constants.

HCHO is both primarily emitted and produced photochemically from the oxidation of VOCs (Friedfeld et al., 2002; Garcia et al., 2006). HCHO photolysis through the radical channel is a daytime source of HO_2 radicals, which recycle to OH upon reaction with NO.

R 1.16	HCHO	+	hv (λ < 325 nm)			$\rightarrow$	H	+	HCO
R 1.17	H	+	O_2	+	M	$\rightarrow$	HO_2	+	M
R 1.18	HCO	+	O_2			$\rightarrow$	HO_2	+	CO
R 1.19	HO_2	+	NO			$\rightarrow$	OH	+	NO_2

Unlike higher aldehydes (see below), the OH reaction with HCHO leads also to the formation of a formyl radical (HCO), which ultimately form HO_2 (see above, R 1.18).

R 1.20	HCHO	+	OH	$\rightarrow$	H_2O	+	HCO

In contrast to other OH radical sources, alkenes ozonolysis is active during the night as well as during the day (Paulson and Orlando, 1996; Rickard et al., 1999; Johnson and Marston, 2008). Rate constants for ozone reaction with alkenes are typically many orders of magnitude smaller than those for the corresponding OH reactions. However, because of the high tropospheric ozone concentration, the contribution of its reaction with alkenes ranges from a few percent to the dominant loss pathway, depending on the structure of the alkenes and local conditions (Paulson and Orlando, 1996). The initial step in the reaction is the addition of O_3 across the double bond to form a primary ozonide, or malozonide. The unstable primary ozonide cleaves simultaneously to give an aldehyde or ketone and an excited intermediate called the Criegee intermediate (after the German scientist, *Rudolf Criegee*, who originally proposed this mechanism).

R 1.21

$$O_3 + \underset{R_2\;\;\;R_4}{\overset{R_1\;\;\;R_3}{\diagdown C=C \diagup}} \longrightarrow \text{primary ozonide} $$

$$R_1R_2C=O + R_3R_4\overset{(+)}{C}OO^{(-)} \qquad R_3R_4C=O + R_1R_2\overset{(+)}{C}OO^{(-)}$$
$$\text{Criegee intermediate} \qquad\qquad \text{Criegee intermediate}$$

The Criegee intermediate contains excess energy and either can be stabilized or decompose in a variety of ways. For example, for the two possible Criegee intermediates produced in the O_3-propene reaction, the following reactions are possible (the numbers above the reaction arrows define the product yields of the corresponding reactions, Finlayson-Pitts and Pitts, 2000):

R 1.22	(HC·HOO)* + M	$\xrightarrow{0.37}$	HC·HOO·	+	M
		$\xrightarrow{0.12}$	HCO	+	OH
		$\xrightarrow{0.38}$	CO	+	H_2O
		$\xrightarrow{0.13}$	CO_2	+	H_2
		$\xrightarrow{0}$	CO	+	2H
		$\xrightarrow{0}$	HCOOH		

Of great importance is the production of OH in these reactions. The OH and other radicals formation from the ozonolysis of alkenes was first observed by Saltzman (1958). Since then extensive studies has been ended to an acceptable explanation of the OH formation pathway (Paulson et al., 1999 and references therein). Paulson et al. (1992) and Atkinson et al. (1992) showed that the OH formation varied widely with alkene structure. For some alkenes the yield of OH is about unity, indicating that the OH formation pathway is the dominant reaction channel.

1.1.2 Radical Propagation

As a result of the radical initiation process, new OH radicals are being generated. These new OH radicals initiate the oxidation of CO and VOCs, resulting in production of hydrogen peroxy (HO_2) and other organic peroxy (RO_2) radicals (Finlayson-Pitts and Pitts, 2000):

R 1.23	$OH + CO$	$\rightarrow$	$H + CO_2$
R 1.24	$H + O_2 + M$	$\rightarrow$	$HO_2 + M$
R 1.25	$OH + CH_4$	$\rightarrow$	$CH_3 + H_2O$
R 1.26	$O_2 + CH_3 + M$	$\rightarrow$	$CH_3O_2 + M$

Unlike HCHO, higher aldehydes react with OH to form peroxyacyl radicals (RCO_3), which then react with NO_2 to form peroxyacyl nitrates (PANs). PANs are labile compounds that thermally decompose to regenerate the RCO_3 and NO_2. NO competes with NO_2 and reacts with RCO_3 to form RO_2.

R 1.27	$RCHO + OH$	$\rightarrow$	RCO_3
R 1.28	$RCO_3 + NO_2$	$\rightarrow$	PAN
R 1.29	PAN	$\rightarrow$	$NO_2 + RCO_3$
R 1.30	$RCO_3 + NO$	$\rightarrow$	RO_2

In the presence of NO_x, the major fate of the RO_2 is through their reaction with NO to form alkoxy radical (RO). A second path for the larger RO_2 radicals is the addition of NO followed by isomerisation to form an alkyl nitrate ($RONO_2$):

| R 1.31 | $RO_2 + NO$ | $\rightarrow$ | $RO + NO_2$ |
| R 1.32 | $RO_2 + NO$ | $\rightarrow$ | $RONO_2$ |

Alkoxy radicals may then react with O_2, decompose or isomerise. When isomerisation (for alkoxy radicals with four or more carbon atoms) is not feasible, reaction of alkoxy radical with O_2 to produce HO_2 and carbonyl compound is always significant and usually predominate. HO_2 radicals react primarily with NO to produce OH:

| R 1.33 | $RO + O_2$ | $\rightarrow$ | $HO_2 + $ carbonyls |
| R 1.19 | $HO_2 + NO$ | $\rightarrow$ | $OH + NO_2$ |

It should be also mentioned that oxygenated volatile organic compounds (OVOCs) that result from the hydrocarbons oxidation process, might lead to the production of new RO_2 and HO_2 radicals (e.g., photolysis of carbonyls and dicarbonyls, Emmerson et al., 2005a, 2005b, 2007). The photolysis of these secondary OVOCs is the main initiation source of the intermediate species RO_2 and HO_2.

Without perturbation from peroxy radicals, photodissociation of NO_2 to NO and subsequent regeneration of NO_2 via reaction of NO with O_3 is sufficiently fast that these species are in a dynamic equilibrium.

R 1.34 $NO_2 + h\nu\ (h\nu < 424\ nm) \rightarrow O(^3P) + NO$

R 1.35 $O(^3P) + O_2 + M \rightarrow O_3 + M$

R 1.36 $NO + O_3 \rightarrow NO_2 + 2O_2$

A photochemical steady state, PSS (Leighton, 1961) exists provided the NO-NO_2-O_3 system is isolated from local sources of NO_x and sunlight is constant.

E 1.1 $[O_3] = j_{NO_2} \times [NO_2] / k_{(O_3+NO)} \times [NO]$

Deviations from the PSS occur when additional sources of NO_2 or other sinks for O_3 exist. In the presence of VOC and NO_x, oxidation of hydrocarbons leads to the formation of RO_2 and HO_2. Recycling of these radicals upon their reactions with NO lead to the formation of NO_2 without consuming O_3. Consequently, deviations from the PSS occur. Ozone production under these conditions is known as summer smog. The ozone photochemical formation through hydrocarbon oxidation processes is a non-linear and can be limited by the VOC or NO_x concentrations.

1.1.3   Radical Termination

Under NO_x limited conditions, the self-reactions of HO_2 and RO_2 radicals compete with NO reactions leading to HO_x radical termination and thus inhibiting O_3 formation (Finlayson-Pitts and Pitts, 2000). The self-reaction of HO_2 leads to the formation of hydrogen peroxide. Hydrogen peroxide is removed from the atmosphere by deposition or it may photolyse regenerating OH or it may react itself with OH:

R 1.37 $HO_2 + HO_2 \rightarrow H_2O_2 + O_2$

R 1.38 $H_2O_2 + h\nu \rightarrow 2OH$

Introduction

R 1.39 $H_2O_2 + OH \rightarrow H_2O + HO_2$

HO_2 is also removed by its cross reactions with RO_2 radicals to form organic peroxides (ROOH). Organic peroxides (ROOH) are removed from the atmosphere by deposition or by photolysis regenerating OH.

R 1.40 $HO_2 + RO_2 \rightarrow ROOH + O_2$

R 1.41 $ROOH + h\nu \rightarrow OH$

HO_2 may even lead to the ozone destruction through its reaction with O_3:

R 1.42 $HO_2 + O_3 \rightarrow OH + 2O_2$

The self-reaction of RO_2 or their reaction with other RO_2 radicals is complex and lead to a variety of compounds:

R 1.43 $RO_2 + RO_2 \rightarrow 2RO + O_2$

               $\rightarrow ROH + RCHO + O_2$

               $\rightarrow ROOR + O_2$

Under high NO_x conditions, the dominant radical sink is the formation of nitric acid (HNO_3):

R 1.44 $OH + NO_2 (M) \rightarrow HNO_3 (M)$

1.1.4   OH reactivity

As shown above, OH exhibits high reactivity to many atmospheric species such as CO, NO_x and VOCs. The OH reactivity is equivalent to the reciprocal atmospheric OH lifetime, τ_{OH}:

E 1.2   OH reactivity = $\sum k_{(i+OH)} \times [i]$,

where $k_{(i+OH)}$ is the bi-molecular rate constant for the OH reaction with species i.

A major uncertainty in atmospheric chemistry results from the incomplete knowledge of the number and abundance of reactive components being present in the atmosphere. Besides well-known pollutants like CO and NO_x, a large number of 10^4 - 10^5 different VOC species exists in the troposphere, but less than one hundred VOCs are being measured routinely in field campaigns (Goldstein and Golbally, 2007). Estimating the OH reactivity based solely on the measured VOC species may thus lead to a significant underestimation of the OH reactivity (e.g., Di Carlo et al., 2004; Mao, 2009; Hofzumahaus et al., 2009, Chatani et al., 2009). Missing reactivity has been reported for different environments, including marine, rural and urban sites with measured to calculated reactivity (based on the measured VOC) as high as 3 (Lou et al., 2009).

Due to its high reactivity, the lifetime of OH is typically less than 1 s and consequently OH rapidly reaches photochemical steady state, given by:

E 1.3 $[OH]_{pss}$ = Rate of production / $\sum k_{(i+OH)} \times [i]$

1.2   Tropospheric Modelling Techniques

As aforementioned, the atmospheric oxidation capacity is largely controlled by the presence of free radicals such as OH and NO_3 and other molecular oxidant species such as O_3. The photochemical reactions governing the formation and destruction of these species as well as their oxidation products are numerous, complex and non-linear. A photochemical model incorporates a chemical mechanism code that describes these photochemical reactions, which the model uses to simulate the photochemical species. Chemical codes can be constructed based on different lumping techniques or explicitly. Photochemical species may be simulated based on zero, one, two or three dimensional photochemical box models and may range from local, regional to global scale.

A zero dimensional box model (as the one used in the current study) assumes that the air is continuously well mixed under the planetary boundary layer (PBL). The PBL may collapses in the model during the night and builds up gradually during the day. The boundary layer height is used in the zero-dimensional box models to describe the deposition velocities of the different simulated atmospheric species. The two most widely used chemical mechanisms are the *Master Chemical Mechanism, MCMv3.1* (http://mcm.leeds.ac.uk/MCM/; Jenkin et al., 1997; Saunders et al., 2003; Bloss et al., 2005a and 2005b), a near-explicit chemical mechanism and the *Regional Atmospheric Chemistry Mechanism, RACM* (Stockwell et al., 1997), which is based on lumping techniques. The simulation of photochemical species using any of these chemical mechanisms requires the solution of a large system of non-linear ordinary differential equations (ODEs) constrained by initial values.

Under conditions of large chemical kinetic systems (i.e., the current case), "stiffness" often occurs. Stiff differential equations arise due to the existence of some solutions that change very rapidly compared with other solutions, or some solutions which change very rapidly at some times and slowly at other times. Non-linear ODEs that are known to be stiff can be handled efficiently by the Gear's explicit multi-step method in which different methods or step sizes could be applied to each equation (Gear, 1971).

FACSIMILE solver used in the current study uses this predictor-corrector Gear's method, in which the values of the solution vector at the end of a step are first predicted, and are then corrected to satisfy the differential equations by a few Newton iterations. FACSIMILE does this efficiently, even for large problems, by exploiting the sparse nature of the matrices involved. Using the above described Gear's predictor-corrector (P-C) multi-step method; FACSIMILE can solve extremely stiff problems efficiently.

1.3 Oxidation Capacity

The physical and chemical properties of the atmosphere are influenced by the presence of trace gases like nitrogen oxides (NO_x) and volatile organic compounds (VOCs). The oxidising capacity of the atmosphere determines the rate of their removal (Prinn, 2003), and hence controls the abundance of these trace gases. Understanding the processes and rates by which species are oxidized in the atmosphere is thus crucial to our knowledge of the atmospheric composition of harmful and climate forcing species. The term "oxidation capacity", OC is defined in the current study as the sum of the respective oxidation rates of the molecules Y_i (Y_i = VOCs, CO, CH_4) by the oxidant X (X = OH, O_3, NO_3) (Geyer et al., 2001):

E 1.4 $OC = \sum k_{Yi} [Y_i] [X]$,

where k_{Yi} is the bi-molecular rate constant for the reaction of Y_i with X.

In addition, the concentration of the oxidant species (O_3, OH, NO_3) was also used as important indicators and key measure of the atmospheric oxidation capacity (Liu et al., 1988). However, since the life time of the trace gases is controlled not only by the oxidant concentration but also by its second-order rate constant (k_{Yi}), the method of Geyer et al. (2001) is most suitable to calculate the relative importance of each oxidant in the current study. Otherwise, O_3 (of the highest concentration among the oxidizing species) would be always the most important oxidant. Recently, a comparable method based on the reciprocal of the lifetime ($1/\tau$) of the oxidized species (Y_i) was also used to express the total oxidation capacity (Cheng et al., 2008).

As aforementioned, the hydroxyl radical (OH) is the primary oxidant in the atmosphere, responsible for the oxidation and removal of most of the natural and anthropogenic trace gases. However, photo-oxidation of volatile organic compounds (VOCs) results in the formation of other important radical intermediates, hydroperoxy (HO_2) and organic peroxy (RO_2) radicals. The recycling of these radicals in the presence of sufficient NO_x levels lead to the formation of the harmful pollutant ozone (O_3). In addition, OH oxidation of aldehydes higher than HCHO leads to the formation of peroxyacyl radicals (RCO_3) (R 1.27, sec. 1.1.2) which form peroxyacyl nitrates (PANs) upon reaction with NO_2 (see reaction R 1.28, sec. 1.1.2). PANs are also harmful pollutants, which can play an important role in the radical chemistry as well as in the delayed formation of ozone on both, local and regional scales (Derwent et al., 2005). $\sum RO_2 + RCO_3$ are hereafter collectively referred to as RO_2. Caused by the importance of the OH radical, identification of its sources and sinks in the atmosphere is crucial for the understanding of the tropospheric chemistry and in determining effective control strategies for harmful species.

Primary sources of the OH radical include the photolysis of ozone (R 1.5 - R 1.8), nitrous acid (HONO) (R 1.9) and formaldehyde (HCHO) (R 1.16 - R 1.18) and the reactions of unsaturated hydrocarbons with O_3 (R 1.21), see sec. 1.1.1. Based on measurement data of the above sources, Ren et al. (2003) calculated the relative importance of the OH initiation sources in New York City and estimated HONO photolysis contributed 56 % followed by ozone photolysis (13 %), alkene ozonolysis (10 %), and HCHO photolysis (8 %).

Previous investigation of the urban oxidation capacity are largely conducted during the summer (e.g., George et al., 1999; Holland et al., 2003; Mihelcic et al., 2003; Ren et al., 2003; Heard et al., 2004; Volkamer et al., 2007; Emmerson et al., 2005b; Emmerson et al., 2007) and spring seasons (e.g., Shirley et al., 2006; Sheehy et al., 2008; Dusanter et al., 2009) owing to the high photochemical activity at this time of the year. During the winter season, investigation of the oxidation capacity was reported for the city centre of Birmingham, UK (Emmerson et al., 2005a and b), New York City (Ren at al. 2006) and Tokyo (Kanaya et al., 2007) which results were compared with those obtained during the summer. The measured radical concentrations in the above studies were often compared with those simulated in order to test the current understanding of the atmospheric chemistry.

Interestingly, the urban daytime OH and HO_2 radical budgets have been shown to be better simulated during summer rather than winter, especially for high NO_x environments. Ren et al. (2006) used a box model incorporating the Regional Atmospheric Chemistry Mechanism, (RACM; Stockwell et al., 1997), which is based on the lumping technique to simulate radical budgets in New York during a winter campaign carried out in 2004. Although they obtained a median measured to model ratio of 0.98 for OH, the RACM model significantly underestimated HO_2, both during day and night, with median measured to model ratio of 6.0. Similarly, during the IMPACT campaign in Tokyo the RACM model reproduced wintertime OH well but underestimated the HO_2 by a median factor of 2. However, during the summer, the RACM model generally reproduced the daytime OH and HO_2 reasonably well (Kanaya et al., 2007). During the springtime in Mexico City, Shirley et al. (2006) reported a median measured to model OH ratio of 1.07 during the morning and night and 0.77 during the rush hour using the RACM model. For HO_2, median measured to model ratios of 1.17, 0.79 and 1.27 were determined during the morning rush hour, midday and night, respectively.

Besides lumped mechanisms, the more explicit Master Chemical Mechanism, MCM (http://mcm. leeds.ac.uk/MCM/; Jenkin et al., 1997; Saunders et al., 2003; Bloss et al., 2005a and 2005b) has been used extensively to interpret field measurements, carried out under a variety of conditions, including urban environments (e.g. Mihelcic et al., 2003; Emmerson et al., 2005a, 2005b, 2007). During the BERLIOZ campaign, which took place in Berlin in August 1998 (Mihelcic et al., 2003), the hydroxyl and peroxy radical budgets have been measured and compared to those calculated by a photochemical box model containing the MCM. The modelled OH concentrations were found to be in excellent agreement with the measurements under high-NO_x conditions (NO_x >10 ppbv). The measured RO_2/HO_2 ratio was also well reproduced by the model. Using the MCM model radical concentrations were also simulated during the TORCH campaign, which took place ~40 km NE of central London in the summer of 2003. Modelled data also agreed well with measurements with only a 24 % and 7 % over prediction for OH and HO_2, respectively (Emmerson et al., 2007).

During the majority of the summer campaign studies reported in the literature the daytime peak OH was well simulated, in the range of $(3-10) \times 10^6$ molecule cm^{-3} (Kanaya et al., 2007 and references therein). However, model OH production rate analysis has suffered from high uncertainties due to the use of estimated HONO concentrations rather than accurate direct simultaneous measurements (e.g. Heard et al., 2004; Emmerson et al., 2005b, 2007; Kanaya et al., 2007).

Using an MCM constrained box model with estimated HONO concentrations, the averaged diurnal OH concentrations during the summer of 1999 PUMA field campaigns in the Birmingham City centre was underestimated by a factor of ~2 during the day especially under high NO_x conditions (Emmerson et al., 2005a). This could potentially be due to an underestimation of daytime HONO concentrations from using only known gas phase chemistry (Kleffmann et al., 2005). Thus, other photochemical sources have been proposed and recently identified in the laboratory, e.g. by the photochemical heterogeneous conversion of NO_2 on natural surfaces (George et al., 2005; Stemmler et al., 2006, 2007). The net HONO photolysis (defined as the HONO photolysis rate minus the radical loss rate due to the reaction OH+NO) was not a net source of OH radicals in the atmosphere when the reaction of NO+OH was assumed as the unique HONO source in the Birmingham city centre (Heard et al., 2004). Emmerson et al. (2005a, 2007) incorporate a parameterization for the heterogeneous conversion of NO_2 into HONO on aerosol surfaces in their MCM model. As a result, an increase in the OH concentration by only 0.03 % (Emmerson et al., 2005a) and a net contribution of HONO to the radical production of 3 % during hot and stagnant "heat wave" conditions of the TORCH campaign (Emmerson et al., 2007) were calculated. A similar contribution of 3 % was estimated in Tokyo assuming heterogeneous production of HONO by dry deposition of NO_2 to the ground with HONO subsequently produced according to the reaction $2NO_2 + H_2O \rightarrow HONO + HNO_3$ (Kanaya et al., 2007). During the LAFRE campaign in California, 1993 (George et al., 1999), a significant reduction in both modelled OH and HO_2 has been observed when heterogeneous HONO formation on ground surfaces was removed from the model, especially in the morning. Similarly, in BERLIOZ, the RACM model predicted only 50 % of the measured OH concentrations when HONO photolysis was switched off in the early morning (Alicke et al., 2003). It is clear, therefore, that the simultaneous measurement of HONO, along side other major radical precursors, is crucial in the analysis of atmospheric radical budgets (e.g. Ren et al., 2003; Kleffmann et al., 2003; Acker et al., 2006a, 2006b).

During the last five years, detailed flux analyses have been often reported showing the different individual contributions of the primary and secondary radical sources and sinks (Emmerson et al., 2005b, 2007; Sheehy et al., 2008; Dusanter et al., 2009). However, none of these studies investigated the balance between the secondary loss (owing to oxidation of CO and VOCs) and production of OH (mainly due to $HO_2 + NO \rightarrow OH$) and its impact on the OH radical initiation rates. Different studies reported similar OH levels measured under different atmospheric conditions of different HO_2 and RO_2 levels or *vice versa*. During the summer PUMA campaign, Emmerson et al. (2005a) observed lower HO_2 concentrations on June 24 when the air arrived from the city centre of Birmingham (NO_x range: 15-30 ppb) than on June 23 when the site experienced westerly airflow (NO_x range: 2-10 ppb) while OH concentrations was the same for both days. Similar results can be obtained from the observation of Emmerson et al. (2007) who found that OH levels were not elevated despite the significantly higher modelled HO_2 and RO_2 levels and the measured RO_2 during the "heat wave" period in comparison to the rest of the TORCH campaign. Similarly, for summer and winter PUMA campaigns, the daytime OH levels in winter were a factor of 2 less than in summer while noontime HO_2 were similar (Heard et al., 2004). These observations imply that RO_2 and HO_2 recycling do not have a significant impact on the observed OH levels. These results however, may contradict with the modelling results of Volkamer et al. (2007) who showed that initiation rates of RO_2 and HO_2 contribute together ~60 % to the total OH initiation rates. A detailed analysis of the balance between the secondary radical production and destruction is presented in sec. 3.2.5.

1.4 Photochemical Ozone Sensitivity

The relation between O_3 and its major anthropogenic precursors, VOCs and NO_x represents one of the major scientific challenges associated with urban air pollution and have been under investigation since two decades of research, effectively after the report of the National Research Council (NRC, 1991). Since then, there have been major advances in the understanding of the photochemical processes covering ozone formation, owed in part to developments in the applied field measurement techniques and sophisticated photochemical models. Ozone is a major environmental concern because of its adverse impacts on human health and on crops and forest ecosystems. In addition, ozone is a major greenhouse gas, and changes in emissions of ozone precursors affect the global climate through a linked chemical system (West et al., 2009 and references therein). Because ozone forms most rapidly in conditions with warm temperatures and sunshine, cities with warm climates are especially likely to experience high ozone events (Sillman, 1999). In each of these locations it is necessary to understand how ozone depends on NO_x and VOC to develop an effective policy response (Sillman, 1999).

The relationship between O_3, NO_x and VOC is complex and non-linear and the knowledge of the controlling factors of this process is essential for understanding regional photochemistry and consequently for applying cost effective control strategies. In addition to its importance for policy, the relation between ozone, NO_x and VOC is worth of attention as a purely scientific problem. The process of ozone formation provides a case study of the interaction between non-linear chemistry and dynamics in the atmosphere, and frequently calls for sophisticated modelling treatment (Sillman, 1999).

Photochemical ozone formation depends on the relative abundances of both, VOC and NO_x. The VOC/NO_x ratio was first identified by Haagen-Smit (1954) as part of the earliest investigation of the ozone formation process. Since then, the impact of the VOC/NO_x ratio has been demonstrated in model calculations and in smog chamber experiments (NRC, 1991). However, the VOC/NO_x ratio does not account for the impact of VOC reactivity, biogenic hydrocarbons, geographic variations or the severity of the events and has been repeatedly shown to fail in more sophisticated photochemical models (Sillman et al., 1999 and references therein). In general, determining the ozone formation sensitivity based solely on the VOC/NO_x ratio without considering the individual VOC reactivities is misleading because total VOC concentrations are dominated by the less reactive alkanes (Sillman et al., 1999). Thus, the determination of the ozone sensitivity requires a photochemical sensitivity analysis that takes into account the VOC reactivity.

Ranking the different VOCs based on their contribution to ozone formation requires a valid reactivity scale that considers the kinetic and mechanistic characters of the VOC oxidation process in the atmosphere. Reactivity scales currently used to estimate hydrocarbon reactivity towards ozone formation include the maximum incremental reactivity, MIR (Carter,

Introduction

1994) and the photochemical ozone creation potential, POCP (Derwent et al., 1996, 1998, 2007a). The MIR and POCP scales take into account the kinetic and mechanistic properties rather than considering only the initial rate of OH attack as in case of the OH reactivity scale (cf. Finlayson-Pitts and Pitts, 2000). The POCP reactivity scale however, is designed for use under northwestern European conditions and hence should be used only within these limited conditions (Derwent et al., 1996).

Milford et al. (1994), Sillman (1995) and Kleinman (2000) established a photochemical relation between ozone sensitivity and oxidised nitrogen species, NO_z ($NO_z = NO_y - NO_x$). Oxidized nitrogen species are oxidation products that are being formed simultaneously by the same reactions leading to ozone formation. Therefore, these photochemical indicator relationships can be used in order to determine whether the atmospheric environment is NO_x- or VOC-sensitive (Sillman, 1995, Kleinman, 2000). Indicator relationships have been used to analyse ozone photochemical formation in Los Angeles, New York, Nashville and Atlanta (Sillman et al., 1997, 1998). These cities span a range of environments from VOC- limited in Los Angeles to transitional in Nashville to NO_x-limited in Atlanta as determined either from measured ratios or from Eulerian model calculations (Kleinman, 2000).

The formulation of an ozone control strategy requires the knowledge of the response of ozone to changes in emissions of NO_x and VOCs (Kleinman, 2000). These parameters are not properties of an air mass as the indicator species but rather depend on the entire sequence of events from the time the NO_x or VOC was emitted to the time the O_3 is measured. These quantities are usually determined with an emission-based model by simply changing the emission rates and seeing how ozone responds. Emission-based models include emission inventories in order to define the sources of the O_3 precursors. There is however, attraction to determining these quantities directly from field observation, not the least of which is an enforced consistency with actual atmospheric concentrations (Kleinman, 2000). In addition, the development of effective ozone control strategies requires explicit sensitivity analyses based on field measurements and the determination of the contribution of each pollutant to ozone formation. The observation-based model analysis (Kleinman, 2000 and references therein) is based on simulating the photochemical ozone formation using on time-dependent observed concentrations of CO, NO and VOCs. The next steps are to apply perturbations to the NO_x or VOC emissions and see how ozone responds.

Photochemical emission- and observation-based models may range from zero dimensional box models to regional trajectory models and up to inter-continental transport models. These models may be based on simplified or highly explicit chemical mechanisms. Huang et al. (2001) used an urban scale photochemical box model based on the Carbon-Bond Mechanism-IV (CBM-IV) (Cardelino and Chameides, 1995) to investigate the photochemical pollution episodes in Tokyo during summer and winter seasons. The applied box model was proven to be an effective tool for the physical-chemical processes associated with urban air quality. Kleinman et al. (2005) performed a comparative study of ozone production and sensitivity in five US metropolitan areas, namely Nashville, New York City, Philadelphia, Phoenix and Houston, using observed concentrations as inputs to a steady state box model based mainly on the RADM2 chemical mechanism (Stockwell et al., 1990). Ozone production varied from nearly zero to 155 ppb h^{-1} with its sensitivity varied from NO_x-limited in rural areas to VOC- limited near emission sources (Kleinman et al., 2005). Similarly, Zhang et al. (2007) used a photochemical observation box model based on the CBM-IV to calculate the sensitivity of the O_3 production to changes in the concentration of the precursor compounds in Hong Kong. Kanaya et al. (2009) used a photochemical box model based on RACM to investigate the ozone sensitivity and ozone photochemical production over Central East China.

Since tropospheric ozone is also affected by transport, disproportion and deposition processes, chemical transport and trajectory models would better represent the ozone sensitivity on regional scale. However, it should be mentioned that such studies are facing the challenge of using a chemical mechanism that adequately describe the photochemical ozone formation and at the same time can be nested in such highly parameterized models and therefore, needs special computer capabilities. Chemical transport models were used for the determination of ozone sensitivity and production in Mexico City (Lei, et al., 2007, Zavala, et al., 2009, Tie et al., 2009) and during 2008 Beijing Olympics (Wang et al., 2009). Chemical transport models were also used to investigate the effect of the intercontinental transport of pollutant emissions on the surface ozone levels in the US (Reidmiller et al., 2009). Similarly,

West et al. (2009) investigated the influences of reduction policies of the ozone precursors on the ozone surface levels between continents and found that the strongest inter-regional influences are from Europe to the former Soviet Union, East Asia to southeast Asia, and Europe to Africa.

It should be also mentioned that the selection between the different modelling techniques depends also on the specific questions that have to be answered. Box models based on explicit chemical mechanisms are most suitable for investigating the ozone sensitivity and ozone photochemical formation explicitly and can provide more information about the secondary processes controlling ozone formation. Derwent et al., (2005) used a photochemical trajectory model with a master chemical mechanism, MCMv3.1 and a highly specified emission inventory to characterize the multi-day ozone formation in northwest Europe. Multi-day ozone formation was shown to be straightforward presented with MCM while it is difficult to represent with the highly simplified chemical mechanisms used in some policy applications (Derwent et al., 2005). Therefore, if the question is purely scientific and the photochemical process has to be elucidated and assessed based on field measurements and state of art knowledge of atmospheric chemistry, then explicit chemical mechanisms should be used in order to extract as much information as possible. Due to the limited computer capabilities, these explicit mechanisms may not be incorporated in 3-dimensional models. However, 3-dimensional and trajectory model based on reduced chemical mechanisms are best suitable for the evaluation of the different emission reduction policies.

Policies to reduce pollutant emissions require also the determination of the emission indices (EI) of these pollutants and identifying their potential sources. The emission index is defined by the mass of the pollutant (in g) per mass of burnt fuel (in kg), which is a very useful number to estimate integrated emissions, if the fuel consumption is known. Different methods are currently used for the determination of the emission indices. Each of these methods encounters certain uncertainties for the determined EI. In addition the application of each of these methods is largely based on the main question to be answered. Emission indices can be determined for individual sources based on vehicle test beds or chassis-dynamometers, a technique, which requires a well-equipped laboratory and is based on simulated driving cycles. However, this technique does not necessarily reflect the real on-road world driving conditions and the level of the actual fleet, expensive and time consuming with low vehicle through output (Oanh et al., 2008 and references therein). For average combustion sources, EI can be determined based on tunnel measurements. Tunnel measurements provide real-world emission indices of vehicles (Sjodin et al., 1998; Stemmler et al., 2005) but, first of all, a well-defined tunnel is required, which may not always be easy to find in many cities (Oanh et al., 2008). In addition, emission indices calculated based on tunnel measurements have the disadvantage of representing only one engine-load condition (i.e., at the nominated vehicle speed in the tunnel). Roadside or rush hour ambient measurements are alternative methods for the determination of the emission indices under real world condition (Oanh et al., 2008) and require simultaneous speciated measurements of NMHCs and NO_x emissions. EI determined based on ambient rush hour measurements have the advantage of representing different engine-load conditions.

1.5 Atmospheric Chemistry Studies in Santiago de Chile

The Chilean anthem has a part that says "Pure, Chile, is thy blue sky, pure breezes flow through thee too…", but this seems to be no longer true because of the poor air quality in Santiago. In Santiago, the average meteorological conditions are unfavourable for the dispersion of air pollutants in the basin, especially during fall and winter (Aceituno, 1988). These stagnant anticyclonic conditions are further intensified in fall and winter by the presence of sub-synoptic features (Rutllant and Garreaud, 1995). The combination of these meteorological conditions and the existence of large emissions of pollutants to the atmosphere determines the occurrence of high concentrations of particles and gases, which have detrimental effects on air quality and visibility (e.g., Trier and Firingueti, 1994; Rutllant and Garreaud, 1995). Maximum concentrations of particles and pollutants, mainly associated with combustion processes, occur in fall and winter months, while in spring and summer there is a great deal of photochemical pollutants (e.g., ozone and PAN) as the actinic fluxes increase (Gallardo et al., 2002).

In view of this unique geographical location and meteorological patterns in Santiago, it is not surprising that earlier studies on air pollution in Santiago, back to the early 1990 were focused on the characterization of aerosol and fine particles (e.g., Horvath and Trier, 1993,

Gramsch et al., 2000, 2004, 2006, 2009; Carvacho et al., 2004, Sax et al., 2007, Richter et al., 2007, Seguel, et al., 2009). Particle concentration levels in Santiago are one of the highest in South America (Gramsch et al., 2006). The coarse particulate matter can reach up to 500 µg m^{-3} on days with strong inversion (Jorquera et al., 1998a). Fine particle ($PM_{2.5}$), elemental and organic carbon and CO levels are also quite high (Gramsch et al. 2000, 2004). Gramsch et al. (2009) showed also that aerosol dynamics play a more important role for Santiago as compared to cleaner cities in Europe. Changes in particle size during different hours of the day reflect both, variations in meteorological mixing conditions as well as effects of aerosol dynamics processes such as coagulation, condensation and dry deposition (Gramsch et al. 2009).

Atmospheric concentrations of polycyclic aromatic hydrocarbons adsorbed on respirable particulate matter and their health effect were studied extensively since around 1999 (e.g., Kavouras et al., 1999; Adonis and Gil, 2000; Gil et al., 2000; Sienra et al., 2005). Recent results from these investigations showed that the major sources of respirable organic aerosol PM_{10} in Santiago are mobile and stationary ones (Sienra et al., 2005). Sienra (2006) investigated also oxygenated polycyclic aromatic hydrocarbons and found that the concentration of these carcinogenic compounds were higher than other cities in both, winter and summer seasons, indicating an exposure of the inhabitants of Santiago to high amounts of carcinogenic and mutagenic air pollutants.

Other studies were focused on the atmospheric concentrations of VOCs, their sources and their reactivities towards ozone formation in addition to PAN and ozone levels (e.g., Rappenglück et al., 2000, 2005; Chen et al., 2001; Monod et al., 2001; Rubio et al., 2004, 2006a, 2007). Of these studies, Monod et al. (2001) and Rappenglück et al. (2000) used different key hydrocarbon ratios in order to have some insights about the hydrocarbon sources and reactivities and their contribution to the photochemical smog formation. Chen et al. (2001) and Rappenglück et al. (2005) found that the leakage of liquefied petroleum gas (LPG) has a significant impact on Santiago air quality. Rappenglück et al. (2005) found that alkanes represent the largest VOC fractions followed by aromatics and alkenes. They found also that alkenes are the most reactive compounds while alkanes do not contribute more than 20% of the total air mass reactivity in Santiago (Rappenglück et al., 2005). Traffic emissions were also shown to be responsible for the formation of secondary organic aerosol (Rappenglück et al., 2005).

Gas phase and dew water incorporated concentrations of hydrogen peroxides have been investigated by Rubio et al. (2006b). Hydrogen peroxide concentrations found in dew were lower than that calculated from the gas phase concentrations based on Henry's law, suggesting a lack of equilibrium (Rubio et al., 2006b). Nitrite concentrations in rain and dew were also investigated in Santiago by Rubio et al. (2002, 2008). From the high nitrite concentrations found in dew, Rubio et al (2002) expected that nitrite could re-suspend in the boundary layer after dew evaporation. Rubio et al. (2002) concluded also that this upward flux could constitute an important source of hydroxyl radicals in the early morning, thus contributing to the photochemical smog formation. However, since gas phase nitrous acid concentrations were not measured in their study, these results were unproved. Recently, Rubio et al. (2009a) measured HONO both, in gas phase and dew and showed that HONO spikes during early morning are more associated to rush hours traffic than to dew re-evaporation.

Several attempts were made to investigate of the formation of the summer photochemical smog as early as in the year 2000. Rappenglück et al. (2000) evaluated the photochemical smog in Santiago de Chile based on a set of ancillary measurements including O_3, CO, NO_x, PAN, C_4-C_{12} NMHCs and meteorological parameters. Using a simple photostationary state approach, Rappenglück et al. (2000) calculated a mean maximum diurnal OH concentrations of ~ $2.9 \cdot \times 10^6$ molecules cm^{-3}. Similarly, in another attempt, Rubio et al. (2005) determined a maximum diurnal OH concentration of ~$8.8 \cdot \times 10^6$ molecules cm^{-3} based also on a simple photostationary state and a set of ancillary measurements including HCHO, O_3, NO_2, PAN and HONO in addition to meteorological parameters. Not that in the first study (Rappenglück et al., 2000), HONO was not measured while in the second (Rubio et al., 2005) alkenes were not considered.

Different modelling efforts were also paid towards the prediction of ozone episodic days and for the assessing of the air quality of downtown Santiago de Chile. Jorquera et al., (1998a) developed a forecast modelling technique for the ozone daily maximum levels at Santiago. This model was based on temperature time series analysis and did not consider any

VOC parameterization because no VOC measurements were available at this time in Santiago. The model therefore was not able to asses the relative importance of the VOCs and NO_x in ozone photochemistry. In addition, the model was suffering from the prediction of some false positive events. In addition, the model forecasted not all episodes. Different box modelling approaches were also developed by Jorquera (2002a, 2002b) in order to assess the contribution of different economic activities to air pollution in Santiago. The model explicitly described the seasonal behaviour of meteorological variables. The model results showed the dispersion condition in fall and winter seasons are 20-30% of the summertime values. Receptor modelling of ambient VOC at Santiago de Chile was also performed by Jorquera and Rappenglück (2004). Based on this analysis, the following source apportionment estimates were obtained: fuel evaporation, 29.7±5.6 %; gasoline exhaust; 22.0±3.4 %; diesel exhaust; 18.1±2.9 %; biogenic, LPG and evaporative emissions, 18.0±3.4 % (Jorquera and Rappenglück, 2004). Schmitz (2005) developed and applied a three dimensional dispersion transport model in order to investigate the air pollution dispersion during summer in Santiago. The model was generally able to reproduce the main surface features of the spatial and temporal variations of the daytime thermally driven circulations in the Santiago basin. The model results showed also that during summer at daytime, most of the contaminants are transported towards the northeast of the city into the Mapocho valley and ventilated out of the basin by up slope winds over the Andes Mountains (Schmitz, 2005). No local or regional photochemical modelling was performed in Santiago so far.

In addition, several studies investigated the general pollution trends and their meteorology dependence (e.g., Rutllant and Garreaud, 1995, Jorquera, et al., 2000, Jorquera, et al., 2004, Schmitz, 2005, Gramsch et al., 2006). Jorquera et al. (2000) used an intervention analysis to determine the seasonal average ozone production rate and found that this parameter shows a steady decrease from 1992 to fall 1995 but from then on shows no clear trend and concluded that the photochemical pollution in Santiago has not improved since 1996. However, because the composition of VOCs has not been systematically measured in Santiago, Jorquera et al. (2000) were not able to determine how much of the earlier ozone trend is related to each precursor emission pattern.

Introduction

Santiago frequently experiences pollution episodes with very high ozone concentrations; often exceeding internationally accepted air quality standards set to protect human health (WHO, 2005; Rappenglück et al., 2000, Rubio et al., 2004). Health problems associated with air pollution in Santiago include increased rates of daily mortality and hospital admissions for respiratory illnesses (Ilabaca et al., 1999; Sax et al., 2007). Cakmak et al. (2007) found a direct positive relationship between peak ozone concentration and daily mortality in Santiago. Similar risks are posed to the population of other major cities in central Chile (Kavouras et al., 2001), namely, Valparaíso, Viña del Mar and Rancagua and in the surroundings of large pollution sources such as the copper smelters located in that area. A risk assessment study demonstrated that there may be large impacts on vegetation and agriculture in central Chile due to the observed high levels of ozone and sulphur dioxide (García-Huidobro et al., 2001). These problems have led authorities to enforce attainment plans for point sources, mainly copper smelters, and the metropolitan area of Santiago.

Chilean environmental authorities have developed a number of control strategies to reduce the potential health effects associated with elevated levels of PM_{10} and ozone since 1994 by passing the "environment-base law" which was completed in 1997 providing a framework for the decontamination efforts in Santiago. In 1995, five monitoring station were established within wider monitoring networks for assessing the air quality of Santiago. The government formally acknowledged the air quality problems in 1996, declaring the Santiago metropolitan region a non-attainment area for ozone, PM_{10}, TSP, and CO. The declaration prompted the environmental commission (CONAMA) to develop an air quality and clean-up plan (Schreifels, 2008 and references therein). In addition, the CONAMA authority of Santiago has implemented a major restructuring of the public transportation sector in February 2007 in order to reduce NO_x and PM_{10} emissions from mobile sources (Schreifel, 2008). Future policy strategies of the environment commission of the metropolitan area of Santiago (CONAMA) include the reduction of the NO_x emissions from stationary sources by 50 % by May 2010.

1.6 Objectives

The main objectives of the present work were the investigation of the oxidation capacity and ozone photochemical formation in the highly polluted city of Santiago de Chile. For this purpose, two field measurement campaigns were performed during summer and winter in the urban area of Santiago de Chile.

The oxidation capacity has been evaluated based on two different modelling approaches, a simple photostationary state (PSS) model incorporating all OH net radical sources and sinks and a zero-dimensional box model based on the Master Chemical Mechanism (MCMv3.1). The MCM-box model has been used to investigate and understand the different photochemical processes governing the HO_x chemistry in such highly polluted conditions. Source apportionment analyses of the net radical sources (i.e., HONO, HCHO, O_3 and alkene ozonolysis) have been performed in order to identify their potential sources. The results of the above analysis have been compared with other studies in order to reach a better understanding of the urban photochemistry.

The sensitivity of the photochemical regime to changes in VOC and NO_x was also investigated as it has a direct influence on the recycling processes of the peroxy radicals. The ozone sensitivity has been studied by simulating the ozone photochemical formation at different VOC/NO_x regimes using a zero-dimensional box model based on the MCM. The results of the model sensitivity analysis have been verified using a set of potential empirical indicator relationship. Ozone production efficiency and the net instantaneous photochemical production rate of ozone have been also determined in order to reach a better understanding of ozone formation. Since summertime photochemical ozone is a sever problem in the city of Santiago, a specifically designed reactivity scale has been also calculated in order to determine the contribution of the different hydrocarbons to ozone formation. The MCM has been further used to elucidate potential factors contributing to the ozone diurnal structure. In view of the role of nitrous acid as an important radical precursor, its contribution to ozone photochemical formation has been also studied. Finally, a local scale ozone control strategy has been determined based on the above sensitivity analysis and the potential of the different NMHCs and has been compared and contrasted to the current applied strategies.

2 Methodology

2.1 Measurements Site

The city of Santiago de Chile (33° 26'S, 70° 40'W, ~600 m above sea level) with a population of almost 6 million is situated in a valley between two large mountain ranges, the Andes (4500 m altitude on average) to the east and the Coastal (1500 m altitude on average) to the west (Fig. 6.1.1, sec. 6.1, p. 137) and is separated by 100 km from the coast of the Pacific Ocean (Gramsch et al., 2006). Owing to this unique geographic location, ventilation and dispersion of air pollution is highly restricted. During the summer months, central Chile is generally under the influence of the sub-tropic anticyclone in the south-eastern pacific, resulting in clear sky and high temperatures in Santiago (Schmitz et al., 2005).

Two measurements campaigns were carried out March 8 – 20, 2005 (summertime) and May 25 - June 07, 2005 (wintertime). The measurements were performed at the campus of the University of Santiago de Chile, USACH (Fig. 6.1.2, sec. 6.1, p. 137). Complementary data were obtained from Parc O'Higgins station (POH), 1.8 km southeast the USACH measurement site. The POH site is part of the local urban network for controlling air quality. Both measurement sites are situated in the downtown area of the city of Santiago. Cluster analysis of pollution data in Santiago de Chile has shown that pollution generated over the city is redistributed in four large areas that have similar meteorological and topographical conditions and that both, USACH and POH stations locate in the same sector and experience the same level of pollutants (Gramsch et al., 2006).

2.2 Measurement Techniques

The techniques used to measure the different parameters are listed in Table 2.2.1 with their response times and detection limits. Similar techniques were used during both summer and winter campaigns with only few exceptions outlined below. A detailed description of the experimental procedures for the parameters measured by the author, namely HONO, HCHO and VOCs is presented. Only a brief description is given for the other data which were obtained directly form the guest institute at the University of Santiago de Chile.

Table 2.2.1: Instrumentation used during the Santiago de Chile field campaigns.

species	method	response time	detection limit
HONO	LOPAP-technique (Long-Path-Absorption Photometer)	4 min	3 pptv
HCHO	Hantzsch reaction based instrument, Aero Laser CH_2O analyser (Model AL4001)	3 min	50 pptv
NO	chemiluminescence based analyzer with molybdenum converter (Model TELEDYNE 200 E)	<10 s	400 pptv
NO_2	DOAS-OPSIS optical system	2 min	0.5 ppbv
$O_3{}^a$	short-path UV absorption (λ = 254 nm, Advanced Pollution Instruments Model 400).	10 s	1 ppbv
$O_3{}^b$	UV absorption based monitor (Dasibi Model 1009-Cp)	10 s	1 ppbv
CO^a	IR absorption based monitor (Interscan 4000)	20 s	1 ppbv
CO^c	IR absorption based monitor (Interscan 4000)	20 s	1 ppbv
$CO_2{}^c$	IR absorption based monitor (Pewatron AG, model Carbondio)	10 s	1 ppmv
PAN	GC-ECD (Meteorolgie Consult GmbH)	10 min	25 pptv
$j(NO_2)$, $j(O^1D)$	filter radiometers (Meteorolgie Consult GmbH)	1 min	—
C_3-C_{10} NMHCs	GC-FID analysis (HP Model 6890) following the US Compendium Method TO-17 (EPA)	3 h (day) and 6 h (night).	~40 pptv (4 -77 pptv)

[a] Measured at the Park O'Higgins station (POH) 1.8 km southeast of the USACH site
[b] Used to investigate the ozone interferences during the VOC sampling process.
[c] Instruments used only during the winter campaign at the USACH site.

2.2.1 HONO Measurements

HONO was measured at the USACH site by the sensitive LOPAP (Long Path Absorption Photometer) technique (Heland et al., 2001; Kleffmann et al., 2002). The LOPAP owed its high selectivity to the use of a two-channel system, which corrects for possible interferences. In addition, the use of an external sampling unit helps avoiding heterogeneous HONO formation on the inner wall of the sampling lines. Caused by the measurement principle (see below), the instrument is the most sensitive HONO instrument with a possible detection limit of 0.2 pptv (Kleffmann and Wiesen, 2008).

2.2.1.1 Analytical Method

According to this method, gaseous HONO is being quantitatively trapped in a stripping coil mounted in the external sampling unit. An acidic stripping solution containing sulfanilamide (R1) is used as effluent. The air is continuously sucked into the stripping coil at defined flow rate by the help of a membrane pump and a flow controller. After separating the gas phase by passing through a debubbler, the effluent containing the formed diazonium salt (I) (see reaction E 2.2) is then pumped to the main unit and treated with a solution containing n-(1-naphthyl)-ethylendiamine-dihydrochlorid (R2). A digitally controlled peristaltic pump provides the liquid flow.

E 2.1 $HONO + H^+ \rightleftharpoons NO^+ + H_2O$

E 2.2 $NH_2\text{-}SO_2\text{-}C_6H_4\text{-}NH_2 + NO^+ \longrightarrow NH_2\text{-}SO_2\text{-}C_6H_4\text{-}N_2^+ + H_2O$
                        R1   I

E 2.3 (diazonium salt with $SO_2\text{-}NH_2$) + R2 (naphthyl-$NH_2^+Cl^-$, $NH_3^+Cl^-$) $\longrightarrow$ $NH_2\text{-}SO_2\text{-}C_6H_4\text{-}N{=}N\text{-}$(naphthyl)$\text{-}NH_2^+Cl^-$, $NH_3^+Cl^-$ (II)

The quantitatively formed azo dye compound (II) (see reaction E 2.3) is continuously pumped into the detection unit which is composed of a special Teflon tube (DuPont, AF2400, 0.56 mm internal diameter and 0.1 mm wall thickness) acting as liquid core wave guide for light introduced by glass fibre optics from a white light diode (LUXEON). The light from the long path absorption tubes is transferred again by glass fibre optics to a two-channel minispectrometer (Ocean Optics SD2000).

In the external sampling unit two stripping coils are used in series. This two-channel system allows HONO to be almost completely trapped in the first stripping coil (channel 1) in addition to a small fraction of any interfering species. The second stripping coil (channel 2) collects about the same amount of interfering species but almost no HONO. From the difference of both channels the interference free HONO concentration is determined.

The LOPAP instrument was recently intercompared against the spectroscopic DOAS technique (Differential Optical Absorption Spectroscopy), both in a smog chamber and in the urban atmosphere. Excellent agreement was obtained under daytime photochemical smog conditions (Kleffmann et al., 2006), in contrast to other intercomparison studies (e. g, Appel et al., 1990; Spindler et al., 2003). The excellent agreement can be explained by the active correction of interferences and the use of an external sampling unit, minimising artefacts in sampling lines. Potential heterogeneous HONO formation on the walls of the USACH building on which the sampling unit was fixed (~130 cm distance), was also investigated. No significant variation of the HONO concentration was observed when varying the distance of the sampling unit (20-150 cm) from the wall.

2.2.1.2   Instrument Calibration

The two channels of the LOPAP system were calibrated before and after each measurement campaign and two times during each campaign. The calibration process is based on two-point linear regression analysis (i.e., blank and calibration standard) using a liquid nitrite (NO_2^-) standard diluted in R1. The corresponding mixing ratios (pptv) of the nitrite standard were calculated by considering the applied gas and liquid flow rates which were also measured each time the calibration proceeds. The background (zero) values were determined based on a polynomial regression equation for the blank measurements performed at different time intervals during the measurements. A zero-corrected HONO signal was obtained by subtracting the zero polynomial fit from the measured HONO absorbance signal. A two-point calibration curve was constructed based on the blank (performed after and before the nitrite standard) and nitrite standard values (see Fig. 2.2.1), both obtained from the zero-corrected

HONO signals. The error bar in Fig. 2.2.1 corresponds to 3 times the standard deviation (σ) of the measured signal.

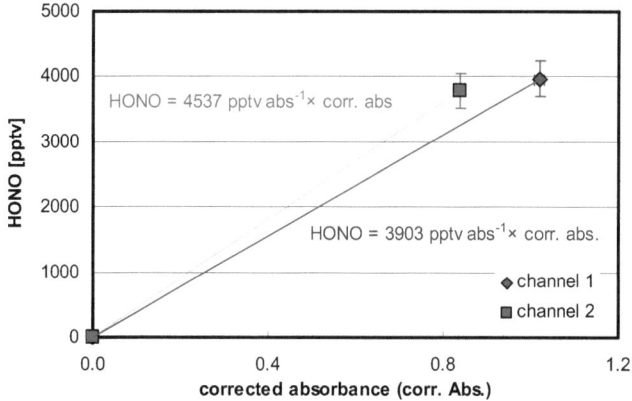

Fig. 2.2.1: Typical two-point calibration curve for the two-channel LOPAP instrument.

2.2.1.3 Instrument Parameters

For the ambient HONO measurements, zero (or blank) measurement were performed on a regular basis (every 8 hours) or whenever required using an automatically regulated air flow system connected to a highly pure nitrogen gas cylinder which acts as zero air source. The two channel HONO concentrations were calculated from the zero-corrected HONO signal (corr. abs.).

The instrument's response time (the time required for the signal to reach 90 % of its maximum value) was 4 min and the time delay (the difference between the measurement time and time when signal appeared) was 11 min. A detection limit of ~3 pptv was obtained during the field measurements which corresponds to 2σ of the measured zero air signals (noise level).

2.2.2 HCHO Measurements

HCHO was measured by a Hantzsch reaction-based Aero-Laser HCHO analyser, model AL4001.

2.2.2.1 Analytical Method

This instrument is based on sensitive wet chemical fluorimetric detection of HCHO (Nasch, 1953), which requires the trapping of the gaseous formaldehyde in aqueous solution before proceeding for analysis. This is achieved by sucking air into a stripping coil fed by the stripping acidic solution using known flow rates. The air and liquid streams are afterwards separated in a glass separator and the solution in then analysed for formaldehyde. A calibrated mass flow controller is used to control the gas flow. A digitally controlled peristaltic pump provides the liquid flow.

The reaction of aqueous HCHO with a solution containing 2,4-pentandione (acetylacetone) and ammonia (NH_3) lead to the formation of 3,5-diacetyl-1,4-dihydrolutidine (DDL), see reaction E 2.4 (Dong and Dasgupta, 1986 and 1987; Dasgupta et al., 1988). The DDL is being excited at 412 nm and the resulting fluorescence at 510 nm is then employed by the instrument for HCHO detection. Studies of interferences showed that this technique is very selective for formaldehyde, with the response for other molecules found in typically polluted air masses being several orders of magnitudes lower (Cárdenas et al., 2000).

E 2.4: $HCHO + 2\ CH_3COCH_2COCH_3 + NH_3 \longrightarrow$ 3,5-diacetyl-1,4-dihydrolutidine

Recently, intercomparison of four different in-situ techniques including the analyzer used in this study, another analyzer employing the Hantzsch technique, FTIR, DOAS and DNPH-sampler was performed (Hak et al., 2005). The results of this study showed that Hantzsch and spectroscopic techniques (FTIR and DOAS) agreed within 15 %. No systematic difference was found between DOAS and Hantzsch instruments under the conditions present during the comparison measurements (Hak et al., 2005).

2.2.2.2 Instrument Calibration

The calibration of the instrument was done before and after each measurement campaign and during the measurements period (e.g., when using new reagents). The calibration procedure is based on two-point calibration method namely blank and standard. The instrument was calibrated using a HCHO liquid standard (*Supelco*, 1.0 ml HCHO, 37 % wt in water). The concentration of the liquid standard was further controlled by iodiometric titration using the method of De Jong (Petersen and Petry, 1985). All solutions were prepared using high purity deionised water (Milli-Q type water) and analytical grade reagents. Time resolution and time delay of the instrument are parameters that depend on the flow-rate settings, therefore, the air- and liquid-flows were also measured each time the calibration proceed. Before the calibration process, the liquid standard was diluted to the required concentration using the same stripping solution connected to the instrument. The standard solution was then sampled to the stripping coil using the same stripping line (instead of the stripping solution) under zero air. The obtained HCHO signals (v) (i.e., blank and standards) were corrected for the background (zero corrected signals) using a polynomial regression equation for the blank measurements performed at different time intervals during the measurements campaign.

A two point calibration curve is calculated based on the signal values of the blank and standard in the zero-corrected data (corrected signal) and using the standard mixing ratio in ppbv (i.e., at the applied air and liquid flows). Fig. 2.2.2 shows a typical two point calibration curve obtained during the measurement campaigns. The error bar in Fig. 2.2.2 corresponds to 3 times the standard deviation (σ) of the measured signal.

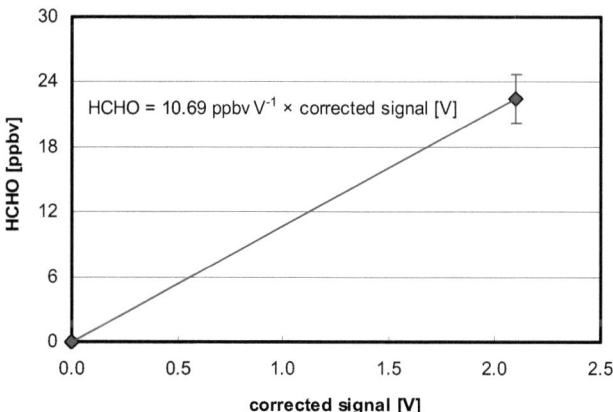

Fig. 2.2.2: Typical two-point calibration curve of HCHO during the field measurements.

2.2.2.3 Instrument Parameters

During the ambient HCHO measurements, zero (or blank) measurement were done every 6 hours or whenever required using an automatically controlled air flow system connected to a highly pure nitrogen gas cylinder for zero air. The ambient HCHO mixing ratios were calculated using the linear fit of the two-point calibration curve and zero corrected HCHO data (corrected signal).

The instrument's response time was 3 min and the time delay 6 min. A detection limit of ~50 pptv was obtained during the field measurements which corresponds to 2σ of the measured zero air signals (noise level).

2.2.3 VOC measurements and analysis

C_3-C_{10} NMHCs were sampled at the USACH site on adsorption tubes and analyzed offline by a GC-FID system following the US EPA Compendium Method TO-17 (Woolfenden and McClenny, 1999). According to this method, the following steps were followed during the calibration process and the analysis of the samples from the field measurements:

Preparation and characterisation of the adsorption tubes

Conditioning of adsorption tubes

Samples collection

Offline GC-FID analysis

Calibration process

Performance criteria

The applied adsorption tubes in the current study are composed of a glass tube (114 mm length and 4 mm i.d., 6 mm o.d.) and a sorbent material. The adsorption glass tube has a 40 mm multi-bed length and is packed with 125 mg Carbotrap graphitized carbon (*Supelco*) followed by 150 mg Carbosieve SIII carbon molecular sieve (*Supelco*) and separated by glass wool plugs. In order to increase the sorbent strength, the Carbotrap has been placed in the front because of its high affinity towards the high molecular weight compounds (C_5-C_{12}) while the Carbosieve SIII further captures the more volatile compounds. This sequence helps improving the sampling efficiency by avoiding the high molecular weight compounds to saturate the strong adsorbent Carbosieve.

The characterization of the adsorption tubes (break through volume, safe sampling volume, storage stability) and the conditioning process were previously described for the same applied method (Niedojadlo, 2005) and were considered accordingly in the current study.

2.2.3.1 Samples Collection

The ambient NMHCs were sampled using an automatic sampling system equipped with calibrated regulated flow controllers and applying regulated air flow of 19 ml/min on the adsorbing tubes. A schematic diagram of the applied sampling system is shown in Fig. 6.2.1, (p.139). Each two adsorption tubes were connected to two different solenoid valves and one flow controller. In order to select which tubes to be loaded at each time intervals, a computer program through a control unit controlled the solenoid valves. During the summer measurement campaign, samples were collected every 3 hours during daytime and 6 hours during night. During the winter campaign, samples were collected at shorter time intervals of 30 min to 1 hour during rush hours, 3 hours during the rest of the daytime and 6 hours during night. Field blank samples were also collected during the measurement campaigns following the same sampling procedure but without connecting them to the sucking lines during the same time intervals. After each measurements campaign, samples were returned to Germany for GC-FID analysis. In the laboratory, the tubes were stored also in a refrigerator under clean environment and were analysed after 1 months storing time at the latest.

2.2.3.2 Offline GC-FID Analysis

Only a brief description of the applied GC-FID system is given here. A detailed description of the analytical procedure can be found elsewhere (Niedojadlo, 2005). The samples have been analyzed by a GC-FID model HP6890 equipped with 90 m HB-1 (nonpolar, 100 % dimethylpolysiloxane) capillary column with 0.32 mm diameter and 3.0 μm film thicknesses. The total GC time run is 80 min which includes 10 min starting temperature of -50°C and then a ramp of 5°C min^{-1} up to 200°C followed by 20 min at constant temperature of 200°C.

2.2.3.3 Calibration Procedure

For the calibration purposes, a custom calibration mixture (RM) was used for the multi-point calibration process. This mixture was first calibrated with a standard gas mixture containing 30 NMHCs (C_2-C_9) compounds prepared and certified by the National Physical Laboratory (NPL), England. A list of the hydrocarbons concentrations in the NPL mixture is presented in Table 6.2.1 (p. 140). The calibration process was performed as follow:

Direct calibration process for the GC-FID system using the NPL standard mixture. This NPL direct calibration process was performed three times.

An individual response factor (RF) was determined for each hydrocarbon in the standard mixture. The response factor is defined as:

$$E\ 2.5 \quad \text{response factor, RF [area g}^{-1}] = \frac{\text{integrated peak area}}{\text{loaded mass [g]}}$$

The calculated individual response factors for the different hydrocarbons in the NPL mixture are shown in Table 6.2.2 (p. 141). For the NPL mixture, an average response factor (Ave-RF) of $1.65 \times 10^{10} \pm 1.53 \times 10^{8}$ area g^{-1} was calculated from the correlation between the average integrated peak area for each hydrocarbon in the three samples and their standard mass concentrations. In order to determine the hydrocarbons mass concentrations in the RM mixture, it was sampled and analysed by the GC-FID system following the same method applied for the NPL mixture. This process was performed also three times and the obtained three chromatograms were accordingly integrated.

The absolute mass of each hydrocarbon in the RM mixture was determined using the Ave-RF determined from the NPL mixture (step 3) and the average integrated peak areas (i.e., for the measured three samples) for the individual hydrocarbons in the RM mixture. A list of the hydrocarbons mass concentration determined for the RM gas mixture is shown in Table 6.2.3 (p. 142) and was calculated as follow:

$$E\ 2.6 \quad \text{mass (RM) [g]} = \frac{\text{average integrated peak area}}{\text{Ave - RF [area/g]}}$$

The multi-point calibration process was performed using an active sampling of the custom made RM mixture over adsorption tubes. The applied adsorption tubes were similar to those used during the field measurements.

After the adsorption tubes thermally desorbed followed by the GC-FID analysis, an individual response factor (area/g) was determined for each hydrocarbon based on a linear regression analysis performed using different concentrations covering the full measurement range. The individually determined RF is henceforth denoted as Ind-RF. The Ind-RF takes into consideration the individual characters of each hydrocarbon that is related to the applied adsorption tubes (see below). An average response factor was also determined and was used to determine the mass concentrations of unknown compounds (i.e., unidentified integrated peaks). In addition to the multipoint calibration process, the RM calibration mixture was also used frequently throughout the measurement period to control the system stability (e.g., retention times), especially after system shutdown.

The calibration curve for α-pinene (see Fig. 2.2.3) is shown as example. A list of the determined Ind-RF values for the other identified hydrocarbons (based on the RM calibration mixture) are shown in Table 6.3.1, Page, 144. For unidentified hydrocarbons, an average RF of $1.35 \times 10^{10} \pm 3.0 \times 10^{9}$ [area/g] was determined from the individual RF determined for the RM mixture. The use of individual RF for each hydrocarbon to calculate the final ambient hydrocarbons concentration rather than the Ave-RF (calculated from the direct calibration with the standard gas cylinder) insures that the individual characteristics of the adsorption tubes due to each individual hydrocarbon are considered.

For the determination of the detection limit (D_l), another set of linear regression analyses were performed for zero concentrations (field blanks) and the very low standards (i.e., ≤10 times the blank level). The standard deviation of the intercept obtained from each of these regression analyses was then used to calculate the D_L (see below).

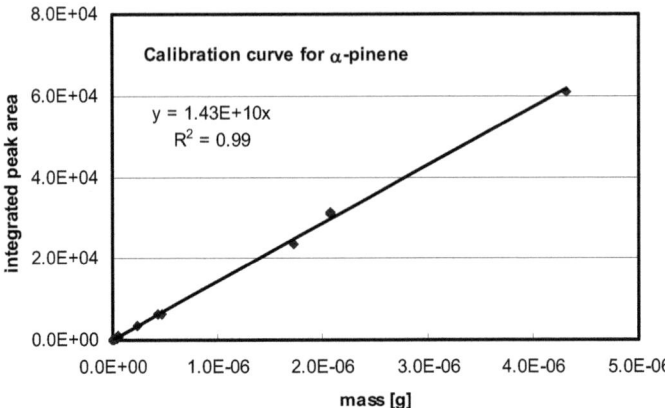

Fig. 2.2.3: Calibration curve for α-pinene performed using the active dynamic sampling of the calibration mixture on adsorption tubes.

2.2.3.4 Performance Criteria

According to the Method-TO-17, the following performance criteria should be met for the applied GC-FID system to qualify for the applied method (Woolfenden and McClenny, 1999):

A method detection limit ≤0.5 ppb.

Accuracy within ±30 %.

Detection Limits (D_L)

The detection limit, D_L has been calculated following the IUPAC procedure of the detection and quantification capabilities no. 18.4.3.7 and was defined accordingly in this study at 95 % confidence limit as:

E 2.7 $D_L = 2 \times 1.645 \times \sigma_o$,

where 1.645 is the student-t value at 95 % confidence limit, (σ_o) is the standard deviation of the intercept (area/g) of the correlation between the integrated peak area and the sampled mass for only the blanks and the very low range of standard concentrations (i.e., ≤ 10 times the blank concentrations). The average and median detection limit calculated for 7000 ml sample was 37 pptv and 36 pptv, respectively (see Table 6.3.1, Page, 144). Since, the D_L values listed in Table 6.3.1, Page, 144 are all within the performance criteria of ≤0.5 ppbv, the applied GC-FID system does qualify for the applied method.

Audit Accuracy

The audit accuracy is defined as the relative difference between the measurement result and the nominal concentration of the nominal standard compound:

$$\text{E 2.8} \quad \text{audit accuracy}, \% = \frac{[\text{nominal concentration} - \text{observed concentration}]}{\text{nominal concentration}} \times 100$$

The nominal concentration is the standard concentration of the RM mixture (see Table 6.2.3, p. 142) loaded on the adsorption tube. The observed concentration is the hydrocarbon concentration determined for the adsorption tube after the thermal desorption and GC-FID analysis. The observed concentrations were calculated for two different sampling volumes (3600 ml and 7480 ml) using the individual response factor (Ind-RF) determined for each hydrocarbon (see Table 6.3.1, Page, 144). In addition, the accuracy was also calculated using the average response factor (Ave-RF) determined from the direct calibration with the NPL mixture. The audit accuracy values calculated for both sampling volumes and using both calculation methods, Ind-RF and Ave-RF, were similar. This result shows that these two volumes lie within the defined range of the safe sampling volume (SSV). As shown in Table 6.2.4 (p. 143), the accuracy values calculated using the Ind-RF lie all within the performance criteria of 30 % while those calculated using the Ave-RF deviate significantly, especially for the important alkene species. Therefore, the applied calibration method based on the individual response factors (Ind-RF) does qualify for the applied method, TO-17.

2.2.3.5 Calculation of Ambient Concentrations

The collected VOC samples during the field campaigns were thermally desorbed and analysed by GC-FID following the same method applied during the RM multi-point calibration process on adsorption tubes. The integrated peak area [relative units] calculated from the GC-chromatogram was then used to calculate the hydrocarbons absolute mass concentrations.

The mass of the different hydrocarbons in each sample were determined from there individual response factors (Ind-RF) determined during the calibration process, see Table 6.3.1 (P. 144) and the integrated peak area of each individual hydrocarbon in the measured sample. For compounds, whose peaks could be integrated but not identified (unknown or unidentified compounds), the average Ind-RF of $1.35\times10^{10}\pm3.0\times10^{9}$ [area/g] was used to calculate the absolute mass. A list of all measured hydrocarbons during the field measurements in Santiago with their Ind-RF are shown in Table 6.2.1 (p. 144). Ind-RF in Table 6.2.1 stands for both, Ind-RF (for known compounds) and the average Ind-RF (for unidentified compounds). In both cases, unknown and identified hydrocarbons, the absolute mass was determined as follow:

$$\text{E 2.9} \quad \text{sampled mass [g]} = \frac{\text{integrated peak area - intercept [area]}}{\text{Ind - RF [area/g]}}$$

The absolute mass can now be converted to ppbv by considering the sampled volume, molecular weight of the hydrocarbon and the average temperature and pressure during the sampling time period. For unidentified hydrocarbons, their molecular weights were calculated as the average molecular weight of the neighbouring identified hydrocarbons in the chromatogram. The relative error that refers to the measured ambient hydrocarbon concentrations was calculated using the method of error propagation. The following equation describes the calculated relative error:

$$\frac{\Delta x}{x} = \sqrt{\left(\frac{\Delta b}{a-b}\right)^2 + \left(\frac{\Delta RF}{RF}\right)^2 + \left(\frac{\Delta FC}{FC}\right)^2 + \left(\frac{\Delta I}{a}\right)^2 + \left(\frac{\Delta NPL}{NPL}\right)^2 + \left(\frac{\Delta RM}{RM}\right)^2 + precision^2}$$

where,

$\left(\frac{\Delta x}{x}\right)$ is the relative error of the measured absolute mass concentration (x).

$\left(\frac{\Delta b}{a-b}\right)$ refers to relative error of the sampled mass and was calculated from the error in intercept given here by the detection limit ($\Delta b = D_L$), integrated peak area (a) and the intercept (b). The value of the D_L was determined based on the error in the intercept obtained from the linear regression analysis of the blanks and the very low range of standard concentrations (see sec. 2.2.3.3).

$\left(\frac{\Delta RF}{RF}\right)$ refers to the relative error of the response factor (RF), where, ΔRF is the standard deviation of the RF of the individual hydrocarbons (see Table 6.3.1, P. 144).

$\left(\dfrac{\Delta FC}{FC}\right)$ refers to the relative error of the used different flow controllers. An average value of 5 % was considered for this parameter caused by the low flows used.

$\left(\dfrac{\Delta I}{a}\right)$ refers to the relative error in integrated peak values, which was given a value of 5 %.

$\left(\dfrac{\Delta NPL}{NPL}\right)$ refers to the relative error in the standard concentrations in the NPL mixture. An average value of 2 % was considered for this parameter (Table 6.2.1, p. 140).

$\left(\dfrac{\Delta RM}{RM}\right)$, refers to the relative error in the standard concentrations of the RM mixture. An average value of 5 % was considered for this parameter.

Precision, refers to the relative error in the calculated integrated peak area for several samples experiencing the same hydrocarbons concentrations. An average value of 10 % was considered for this parameter.

2.2.3.6 Ozone Interferences

Potential ozone interferences have been tested in the laboratory by sampling the standard RM mixture over the same type of adsorption tubes with and without addition of ozone at a mixing ratio of 135 ppbv. Sampling periods of three hours were chosen using NMHCs mixing ratios corresponding to the minimum observed NMHCs mixing ratios during the measurement campaigns.

Ozone was prepared as shown in Fig. 2.2.4 by passing a regulated flow of pure synthetic air through a mercury UV-lamp based ozoniser followed by a reaction vessel with glass rings cooled with dry ice to 203 K in order to trap the HO_x radicals from the ozonised air. Ozone has been monitored by a commercial UV absorption based monitor (Table 2.2.1). Only reductions of as low as -8.8 % for *trans*-2-butene and as high as -29.4 % for *cis*-2-pentene were observed. The average and median O_3 values (averaged over the same time intervals of the VOC samples) during summer were only 26 and 21 ppbv, respectively. Thus, we exclude significant negative interferences from ozone. This result is in agreement with the study of Koppmann et al. (1995) who found no significant interferences from ozone for mixing ratios up to 100 ppbv either using pressurized air samples or cryogenically collected air samples even at very low VOC concentrations.

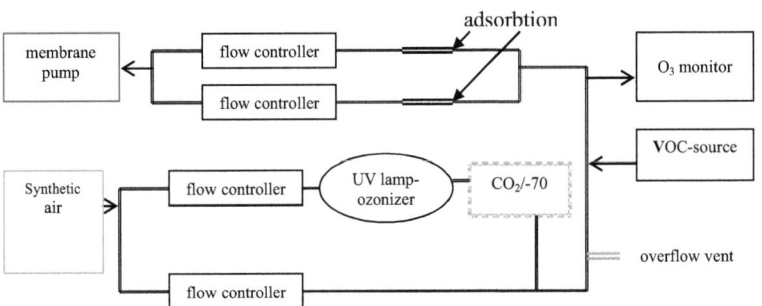

Fig. 2.2.4: Schematic diagram for the active sampling of the VOCs during the ozone interference tests.

2.2.4 Other Parameters

Groups other than the Physical Chemistry Lab of the University of Wuppertal measured the following parameters and the data were obtained directly from the guest institute at the University of Santiago de Chile. Therefore, only a brief description of the deployed techniques is presented.

2.2.4.1 PAN Measurements

PAN was measured at the USACH site by gas chromatography employing a phenylmethyl polysiloxane / dimethylpolysiloxane capillary column and electron capture detection (GC-PAN from Meteorologie Consult GmbH, Germany) with 25 pptv detection limit (Volz-Thomas et al., 2002). The equipment provides quasi real time measurements, at 10 min time intervals. The calibration system is based on the photolysis of acetone in the presence of NO in a flow reactor, with a reaction yield for PAN of 92 ± 2 %. The estimated precision of the PAN measurements was 9 %. The technique has been validated in inter-comparison experiments (Schrimpf et al., 1995).

2.2.4.2 NO_x Measurements

NO was measured at the USACH site by a chemiluminescence based analyzer (Model TELEDYNE 200 E) with a detection limit of 400 pptv. The NO_2 channel of the instrument was not used, since the molybdenum converter converts all reactive nitrogen species (NOy) that may enter the converter such as HNO_3, HONO, N_2O_5, $\Sigma RONO_2$ and $\Sigma PANs$ to NO which are then quantified by the analyzer.

Therefore, NO_2 was measured using an OPSIS-DOAS system with 453 meters path length and a detection limit of 0.5 ppbv.

2.2.4.3 Photolysis Frequencies

Photolysis frequencies $j(NO_2)$ and $j(O^1D)$ were measured by separate filter radiometers (Meteorologie Consult GmbH). The photolysis frequencies of HCHO and HONO were calculated from the measured $j(NO_2)$ and $j(O^1D)$ data using published parameterisation (Holland et al., 2003).

The parameterisation by Holland et al. (2003) covers a variety of meteorological conditions (zenith angle = 31° - 90°, 300 - 400 DU total ozone, 6° - 28°C ambient temperature). In Santiago, the average campaign maximum zenith angle of 32°, annual mean (1979-2004) total ozone column of 281±8 DU (Diaz et al., 2006), clear sky and ambient temperature range of 12° - 32°C, lies within the range of meteorological parameters used to generate the *j*-parameterization by Holland et al., 2003 justifying its use.

2.2.4.4   CO Measurements

During the summer campaign, CO data (IR absorption based monitor, Interscan 4000 monitor) were obtained from POH station. During the winter campaign, carbon monoxide was measured with a similar monitor at the USACH site. The apparatus at the USACH site was calibrated with a reference gas mixture (AGA by Linde Company) containing 10.3±0.26 ppmv CO in purified nitrogen. The monitor provides real time CO concentrations. Measurements were carried out every 20 seconds with a detection limit of 1 ppbv.

2.2.4.5 CO_2 Measurements

Carbon dioxide was measured at the USACH site during only the winter campaign. Carbon dioxide was determined with a Pewaton AG, Carbondio 1000 ppm model monitor. CO_2 data were obtained with ±2 % accuracy in the 0 to 1000 ppmv range (detection limit 1 ppmv). The equipment provides real time measurements and was calibrated with pure nitrogen and a calibrated reference gas mixture (AGA by Linde Company) containing 745±7 ppmv CO_2.

2.2.4.6   Ozone Measurements

O_3 was measured at the POH site with a short-path UV absorption (λ = 254 nm), from Advanced Pollution Instruments Model 400. The instrument provides real time measurements with 10 s response time and 1 ppbv detection limit.

2.2.4.7   Meteorological Parameters

Meteorological parameters such as temperature, relative humidity, pressure, wind speed, wind direction were obtained directly form POH station.

2.3 Modelling Approach

Two modelling approaches, namely a simple quasi-Photo-Stationary State model (PSS) and a zero dimensional photochemical box model based on the Master Chemical Mechanism, (MCMv3.1) were used to analyse the radical budget and concentration during the summer and winter campaign. The ozone sensitivity and the potential of different hydrocarbons to ozone formation were also determined using the MCMv3.1.

2.3.1 Simple Quasi-Photo-Stationary State Model, PSS

OH concentrations were calculated with the steady-state approximation using the radical production rates from HONO, HCHO and ozone photolysis and from alkenes ozonolysis in addition to the net radical loss rate. Under the prevailing high NO_x conditions radical loss is mainly governed by the reactions of OH with NO_x (c.f. George et al., 1999; Ren et al., 2006; Emmerson et al., 2005b, 2007; Kanaya et al., 2007, Dusanter et al., 2009). During the day, formation of HONO by reaction of OH with NO is essentially balanced by photolysis of HONO formed from this reaction. Radical removal by peroxy-peroxy radical reactions is unimportant under high NO_x conditions (see sec. 3.3.3). Thus, the net radical loss rate (L_R) can be simply estimated from the rate of the reaction OH + NO_2:

$$L_R = k_{NO_2+OH}[NO_2][OH]$$

The applied steady state approximation can be summarized as follows:

$$P_R = L_R.$$

The total rate of radical initiation, P_R, is given by:

$$P_R = P_{OH}(prim) - k_{OH+NO}[NO][OH] + P_{HO_2}(prim), \text{ where:}$$

$$P_{OH}(prim) = j(HONO)[HONO] + j(O^1D)[O_3]\Phi_{OH} + \Sigma k_{O_3+alkene}[alkene][O_3]\Phi_{OH},$$

$$P_{HO_2}(prim) = 2j(HCHO_{radical})[HCHO].$$

For ozone photolysis, Φ_{OH} (defined here as the yield of OH radicals formed per ozone molecule photolysed) was calculated using known rate constants (Atkinson et al., 2004) for O^1D quenching and reaction with water in addition to the measured water concentration. For the alkene ozonolysis reactions Φ_{OH} represents the OH yield from the respective reactions (e.g. Rickard et al., 1999). HCHO photolysis is considered as an initiation sources, since

Methodology

during the summer campaign in Santiago, only ~30 % of the measured HCHO has been shown to be photochemically produced (see sec. 3.2.9.1), a fraction which may be even lower during winter as a result of lower photochemical activity.

Therefore, the steady state OH concentration is given by:

[OH]$_{PSS}$ = P$_R$ / (k_{NO_2+OH}[NO$_2$]).

2.3.2 The Master Chemical Mechanism, MCM

A zero dimensional photochemical box model based on the Master Chemical Mechanism, MCMv3.1 (http://mcm.leeds.ac.uk/MCM) has been used to evaluate the radical budgets. MCMv3.1 is a near-explicit chemical mechanism describing the detailed gas phase tropospheric degradation of methane and 135 primary emitted NMHCs, which leads to a mechanism containing *ca.* 5900 species and 13500 reactions. The mechanism is constructed according to a set of rules as defined in the latest mechanism development protocols (Jenkin et al. 1997; Jenkin et al., 2003, Saunders et al., 2003; Bloss et al., 2005a and 2005b). The MCM photochemical box model system of simultaneous stiff ordinary differential equations (ODEs) was integrated with a variable order Gear's differentiation method (FACSIMILE; Curtis and Sweetenham, 1987). Dry deposition terms have been incorporated in the model based on the values of Sommariva et al. (2007) for HNO$_3$ (2 cm s^{-1}), NO$_2$ (0.15 cm s^{-1}), PAN and other PANs (0.2 cm s^{-1}), O$_3$ (0.5 cm s^{-1}), SO$_2$ (0.5 cm s^{-1}), H$_2$O$_2$ (1.1 cm s^{-1}), organic peroxides (0.55 cm s^{-1}), methyl and ethyl nitrate (1.1 cm s^{-1}) and HCHO and all other aldehydes (0.33 cm s^{-1}). In the model, the boundary layer height collapse to 300 m during the night and then gradually builds during the morning to 1.4 km. Dry deposition terms were calculated as V_i/h where V_i is the species dependent dry deposition velocity and h is the (time dependent) boundary layer mixing height.

2.3.2.1 Simulation of Radical budgets

For the modelling of radical levels and their budgets, the model was constrained with campaign average 10 min values of the following measured parameters: j(NO$_2$), j(O^{1}D), relative humidity, pressure, temperature, NO, NO$_2$, HONO, CO, HCHO, O$_3$, PAN and 31 NMHCs (see Table 3.1.1). j(HONO) and j(HCHO$_{radical}$) were parameterised from the measured j(NO$_2$) and j(O^{1}D) (Holland et al., 2003) and their values have been constrained in the model (see sec. 2.2.4.3). The other photolysis frequencies are parameterized within the model using a two stream isotropic scattering model under clear sky summertime conditions (Hayman, 1997; Saunders et al., 2003). The photolysis rates are calculated as a function of solar zenith angle and normalized by a scaling factor, calculated from the ratio of measured and model calculated $j(NO_2)$ values, which takes into account the effects of varying cloud cover and aerosol scattering.

A series of rate of production (ROPA) and destruction (RODA) analyses were carried out in order to identify the most important photochemical processes driving the formation and loss of OH and HO_2. The MCM photochemical model was run for a period of five days, with the model being constrained with the same measured campaign average parameters each day, in order to generate realistic concentrations for the unmeasured intermediate species. By the fifth day the free radicals in the model have reached a photostationary state, which has been used for the data evaluation.

2.3.2.2 Photochemical Simulation of Ozone

The zero dimensional photochemical box model described earlier (see sec. 2.3.2) that is based on the Master Chemical Mechanism, (MCMv3.1: http://mcm.leeds.ac.uk/MCM) has been used to determine the ozone sensitivity and the hydrocarbon potentials to form ozone under typical summertime conditions in Santiago.

The base model was run unconstrained to the measured O_3, NO_2 and PAN for 5-days with the model being constrained with the same measured campaign average parameters each day, in order to generate realistic concentrations for the unmeasured intermediate species. The data obtained at the 5th day base model was compared with those obtained from the 1st day base model, which has been further considered for the different sensitivity analysis (see below). The results of the sensitivity analysis performed based on the 1-day base model were similar to those obtained using the 5-day model run (see sec. 3.3.3). Thus, because the 1-day base model has the advantage of shorter integration time, it has been further considered for the different sensitivity analysis.

For the analysis of the empirical indicator relationships (sec. 3.3.4) and the photochemical ozone production (sec. 3.3.6), the average diurnal concentrations of the species HNO_3, alkyl nitrates ($\Sigma RONO_2$), peroxyacyl nitrates ($\Sigma PANs$), N_2O_5, the radical species OH,

HO_2 and RO_2 and the production rate of alkyl nitrates ($P(RONO_2)$)) were obtained from the results of the 5-day constrained model run used to investigate the summertime oxidizing capacity of Santiago (sec. 2.3.2).

Different model scenarios were used in the present study to determine the O_3-VOC-NO_x sensitivity and the individual hydrocarbons potentials to the photochemical ozone formation:

Scenario_1: Photochemical ozone formation was simulated using the one day base model (see above) that was run unconstrained to the measured O_3, NO_2 and PAN. The PAN was unconstrained from the model in order to allow simulations of O_3 without interference of the modelled NO_2 concentrations by the measured PAN.

Scenario_2: This scenario is designed to determine the ozone sensitivity towards VOC and NO_x from a series of models, each with an incremental factor by which the measured NO or VOC concentration was multiplied.

Scenario_3: For determining the individual hydrocarbon potential to ozone formation, a series of model sensitivity experiments was performed, in each of which the system was perturbed by a small increase of 5 % on molar basis in the VOC of interest. The extra ozone formed compared to scenario_1 ($\Delta O_3/\Delta VOC_i$), averaged over a 24 h time period, was then used to determine an absolute reactivity scale for the measured VOCs to the ozone formation, henceforth called as photochemical incremental reactivity scale, PIR:

$$\text{E 2.10} \quad PIR = \frac{\Delta O_3}{\Delta VOC_i} = \frac{\text{mean}(O_{3i} - O_3\_\text{base model})}{\text{mean}(VOC_i - VOC)}$$

where O_{3i} is the ozone simulated as a result of a 5% increase in the VOC species of interest, VOC_i. The PIR is defined as the number of O_3 molecules formed per molecules of $VOCi$ added.

Scenario_4: The diurnal contribution of the different species (individual VOCs, total VOCs, CO, HCHO and CH_4) to ozone formation was determined from a series of model runs, in each of which the concentration of the species of interest was set to zero.

Scenario_5: The HONO contribution to ozone formation was determined by running the base model unconstrained to the measured HONO, i.e. only photostationary state HONO concentration, $[HONO]_{pss}$ (see sec. 3.2.6) was included.

The contribution of the above species (scenario_4 and 5) to ozone formation was then determined from the reduction of the ozone formation compared to the base model.

Scenario_6: For the ozone control strategy, an additional scenario in which both VOCs and NO levels were reduced by 50 % (scenario_6) was performed.

3 Results and Discussion

3.1 Measurements Results Analysis

For the data evaluation, all measurements were averaged over 10 min time intervals. The measurement data of the trace gases and the meteorological parameters during the summer and winter campaign are shown in Fig. 3.1.1 - Fig. 3.1.5. During the whole summer campaign March 8 – 20, 2005, sunny weather conditions were prevailing. However, during the winter campaign May 25 – June 6, sunny and clear sky conditions were available only during five days of the campaign, from May 29 to June 1 and June 6, 2005. During the rest of the campaign, unstable overcast weather conditions prevailed. Therefore, only the data measured during these clear sky days were considered in the following analysis. The time is reported as local time, which is equivalent to the universal time clock (UTC) - 4 hours during winter and UTC - 3 hours during summer. Therefore, the photolysis frequencies are shifted by 1 h during summer in comparison to that during winter (see Fig. 3.1.4). The daytime is defined from 08:00 to 19:00 h during summer and from 08:00 to 17:00 h during winter.

3.1.1 Summer Campaign

During the summer campaign, the wind speed was relatively low ranging from 0.2 m s^{-1} to 4.1 m s^{-1}, and the average relative humidity was 49 %, reaching up to 100 % during the night with temperatures ranging from 285 K to 305 K during daytime.

The maximum HONO mixing ratio during rush hour reached ~7 ppbv on March 10, 2005 at ~09:00 h (see Fig. 3.1.1). For the campaign averaged data, maximum and minimum HONO mixing ratios of 3.7 ppbv at around 08:00 h and 1.5 ppbv around 17:00 h were obtained.

For CO and NO a similar rush hour peak at ~09:00 h on March 10, 2005 was also observed with maximum concentrations of 3.6 ppmv and 480 ppbv, respectively. The average daytime rush hour maxima for CO and NO were 1.38 ppmv and 180 ppbv, respectively (see Fig. 3.1.4). The NO_2 maximum was shifted later owing to small direct emissions and formation by the reaction of NO with peroxy radicals and O_3. From the slope of the correlation plot of HONO against NO_x a mean HONO/NO_x ratio of 0.008 was estimated during the rush hour peaks, which is in excellent agreement with direct tunnel measurements in Europe (Kurtenbach et al., 2001). PAN, HCHO and O_3 showed typical diurnal variations with average daytime maxima at ~14:00 h of 3 ppbv, 7 ppbv and 65 ppbv, respectively, demonstrating their photochemical formation (see Fig. 3.1.4). However, from the fast increase of HCHO in the early morning, when the O_x ($NO_2 + O_3$) increase was still small, a significant contribution from direct emissions was also identified (see sec. 3.2.9.1). In addition to the maximum at ~14:00 h, the ozone diurnal variation profile is characterized with an afternoon shoulder at 18:00 h (see Fig. 3.1.4), which has become a typical feature under photochemical smog conditions in Santiago (Rappenglück et al., 2000, 2005 and see sec. 3.3.8).

The daytime HONO concentrations are significantly higher than in other polluted urban areas such as New York, Milan or Rome, where the minimum mean daytime concentrations were 0.3 - 0.6 ppbv (Ren et al., 2003; Kleffmann et al., 2006; Acker et al., 2006b). The high mixing ratios of HONO and the daytime maximum of the HONO/NO_x ratio (see Fig. 3.1.4) in Santiago point to a very strong daytime HONO source.

Measurements results analysis

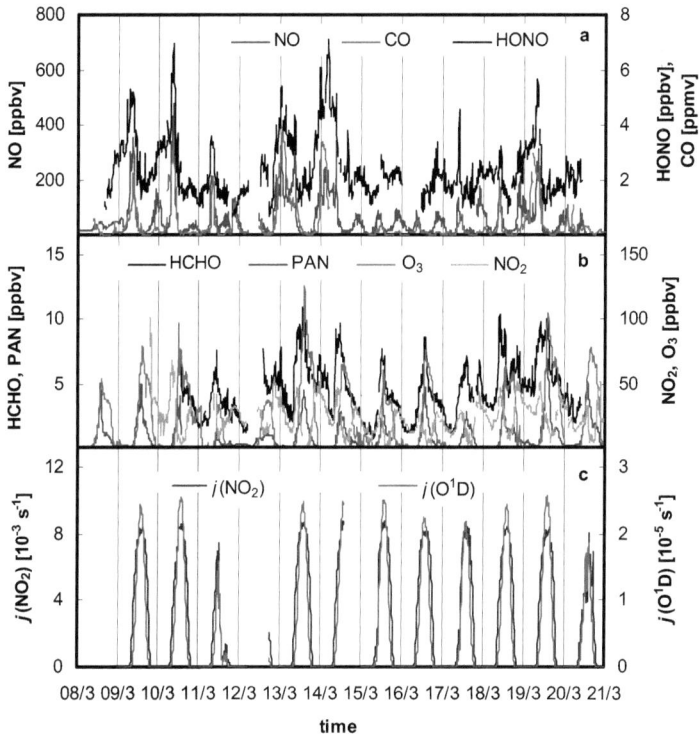

Fig. 3.1.1: 10 min average data of the measured species during the field summer campaign in Santiago de Chile, March 8 - 20, 2005.

During the summer, >200 NMHCs (non-methane hydrocarbons) were quantified (see Table 6.3.1, P. 144), of which only 62 were identified (see Table 3.1.1). Fig. 3.1.2 shows the average diurnal variation of NO and the total quantified NMHCs during summer. The identified NMHCs accounts on average for 43 % of the total quantified NMHCs of ~900 ppbC.

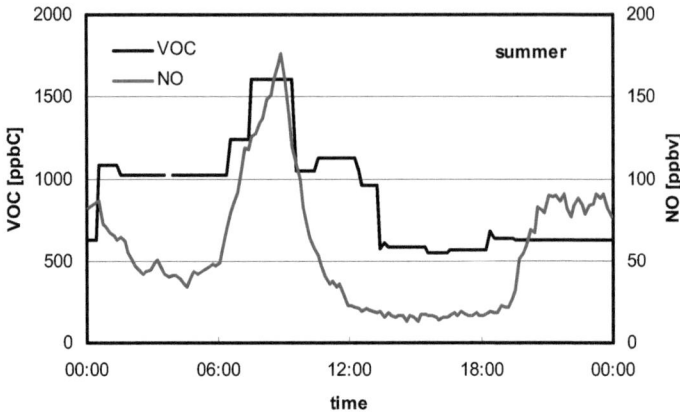

Fig. 3.1.2: Average diurnal variation of the total quantified VOCs and NO during the summer measurement campaign, 2005.

3.1.2 Winter Campaign and Comparison to Summer

The dominant wind direction during daytime in winter was south-west (~225°) while it was from the south-east (~130°) during night. The average wind speed during daytime was ~1.2 m s^{-1} decreases to ~0.6 m s^{-1} during night. Similar wind profiles have been previously reported for Santiago in winter and summer (Gramsch et al., 2006). The average temperature during winter ranged from 290 K during daytime to 280 K during night, while the average relative humidity during daytime was 57 % reaching up to 91 % during the night.

The species diurnal profiles in winter were generally characterized by two maxima, one during the rush hour (~9:00 h) and one late in the night (~22:00 h). The high night time maxima were not so pronounced during summer and can be explained by the more stagnant conditions seen during the winter given by the very low wind speed and the much lower boundary layer height (BLH) during winter (Gramsch et al., 2006). NO, CO and CO_2 show similar diurnal profiles, reaching their maxima of 650 ppbv, 5 ppmv and 657 ppmv, respectively on May 31, 2005 during the morning rush hour while they reach 1113 ppbv, 10.8 ppmv and 861 ppmv, respectively on May 30, 2005 at night (~22:00 h). HONO, NO_2 and HCHO, also show similar diurnal night maxima of 14 ppbv, 144 ppbv and 13 ppbv, respectively on May 30, 2005. O_3 and PAN reach a maximum of 34 and 3 ppbv on May 31 and June 1, respectively during the afternoon at ~16:00 h (see Fig. 3.1.3).

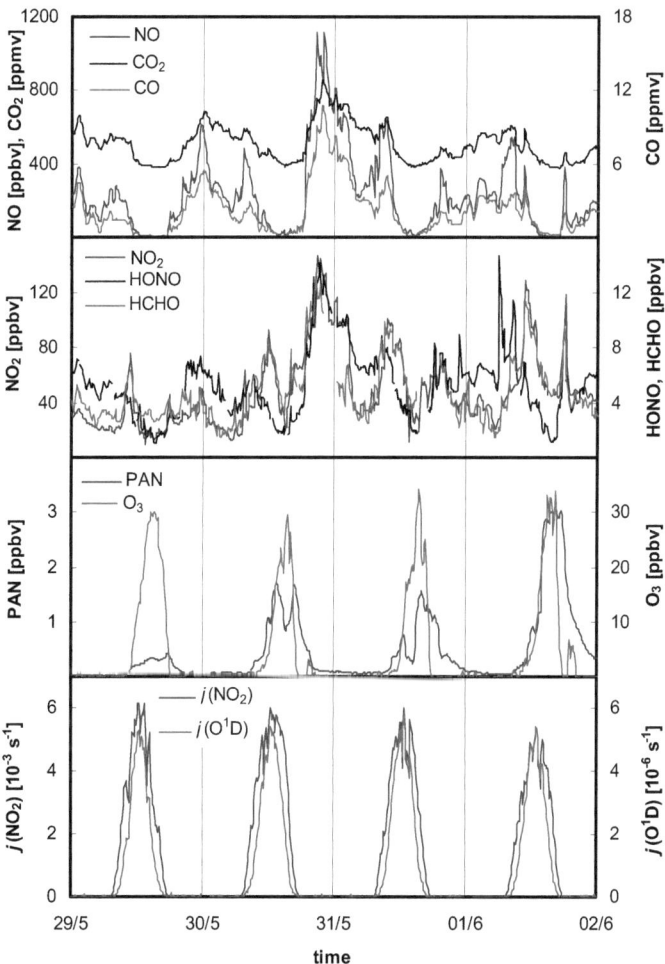

Fig. 3.1.3: 10 min average data of the measured species during the winter field campaign in Santiago de Chile, May 29 – June 2, 2005.

Measurements results analysis

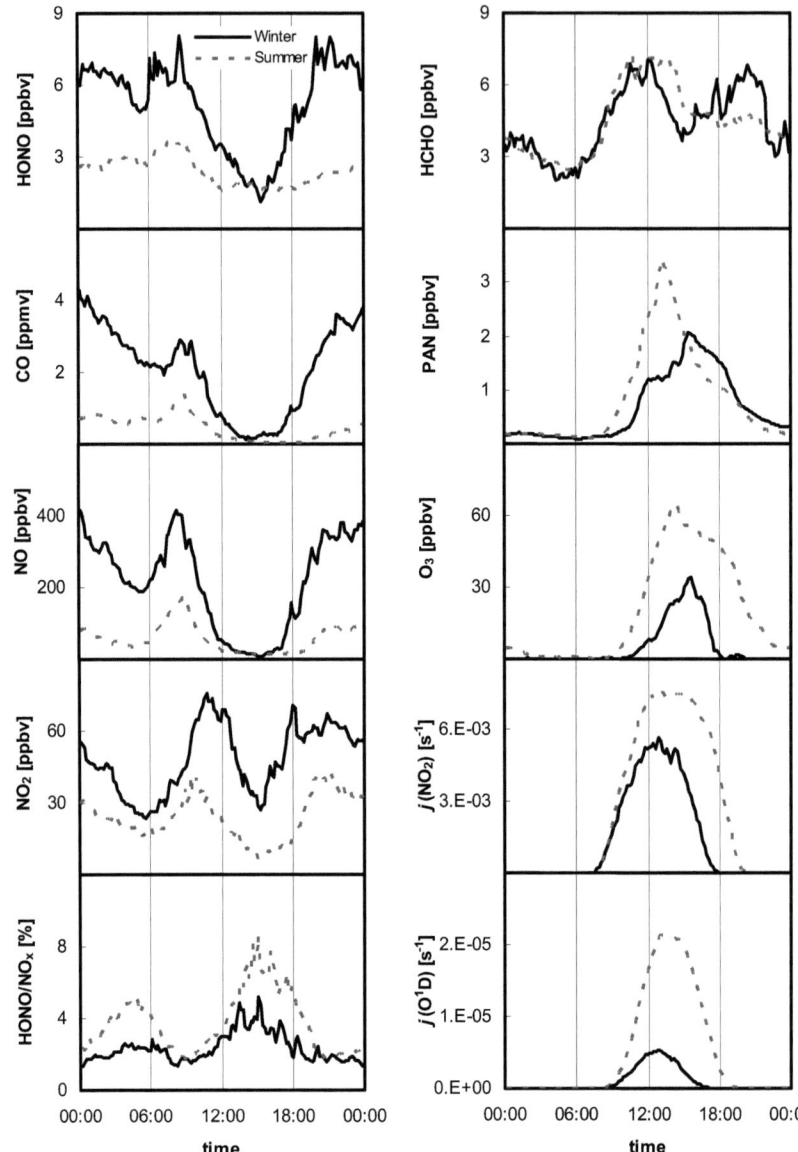

Fig. 3.1.4: Average diurnal variations of the measured species during winter and summer. Due to the use of local time, the photolysis frequencies are shifted by 1 h in summer in comparison to that of winter (see sec. 3.1).

Measurements results analysis

Table 3.1.1: List of the VOCs measured during the summer and winter campaigns.

MCM	compound name	D_L (ppbv)	mixing ratio (ppbv)					
			summer			winter		
			max	average	median	max	average	median
*	propene	0.07	38.8	3.8	1.56	20.1	5.30	4.46
*	propane	0.07	475	41.8	11.5	196	44.0	28.9
*	i-butane (2-methylpropane)	0.06	18	2.95	1.57	26.5	6.07	4.34
	1-butene, i-butene	0.2	9.2	2.35	2.04	10.9	4.71	4.07
*	butadiene	0.06	0.41	0.15	0.12	1.83	0.34	0.29
*	n-butane	0.04	18.3	3.89	2.32	41.0	7.79	5.20
*	trans-2-butene	0.04	0.86	0.18	0.11	2.83	0.46	0.30
*	cis-2-butene	0.04	0.67	0.16	0.1	2.28	0.42	0.31
*	3-methyl-1-butene	0.05	1.22	0.19	0.14	1.21	0.31	0.26
*	i-pentane (2-methylbutane)	0.06	27.6	5.75	4.08	59.6	10.7	7.98
*	1-pentene	0.01	1.77	0.3	0.18	3.40	0.79	0.71
	n-pentane, 2-methyl-1-butene	0.06	18.8	2.03	0.82	30.3	4.81	3.40
*	isoprene	0.02	1.84	0.67	0.51	3.33	0.75	0.57
*	trans-2-pentene	0.01	1.41	0.24	0.14	2.84	0.47	0.32
*	cis-2-pentene	0.004	0.74	0.15	0.1	1.24	0.32	0.28
*	2-methyl-2-butene	0.04	2.1	0.33	0.2	3.63	0.65	0.50
	2,2-dimethylbutane	0.09	5.2	1.07	0.63	9.98	2.70	2.47
	cyclopentene	0.02	0.17	0.05	0.04	0.51	0.09	0.07
	methyl-tert-butyl ether, 2,3-dimethylbutan, cyclopentan	0.16	5.84	1.28	0.89	8.96	2.59	2.30
	2-methylpentane	0.06	14.4	3.11	1.94	24.0	6.80	6.60
*	3-methylpentane	0.05	6.09	1.4	0.92	11.0	3.21	3.08
*	1-hexene	0.02	0.8	0.2	0.14	1.26	0.49	0.50
	n-hexan, 2-ethyl-1-butene	0.06	5.31	1.36	0.94	10.5	3.31	3.13
	2,3-dimethyl-1,3-butadiene	0.02	0.15	0.05	0.03	0.24	0.07	0.07
	methylcyclopentane, 1-methyl-1-cyclopentene	0.07	6.27	1.36	0.87	12.6	3.07	2.66
*	2,3-dimethyl-2-butene	0.02	0.52	0.1	0.06	0.82	0.27	0.26
*	benzene	0.08	9.22	2.13	1.43	22.1	6.89	7.14
	cyclohexane, 2,3-dimethylpentane	0.07	8.07	2.08	1.5	21.5	4.49	3.85
*	2-methylhexane	0.04	1.71	0.35	0.22	2.19	0.60	0.59
	cyclohexene	0.05	0.39	0.13	0.1	2.03	0.42	0.33
	1-heptene	0.02	1.01	0.24	0.17	1.57	0.48	0.48
	2,2,4-trimethylpentane	0.03	3.81	0.92	0.68	4.66	1.61	1.46
*	n-heptane	0.02	2.5	0.55	0.42	4.95	1.24	1.15
	1,4-cyclohexadiene	0.04	0.14	0.07	0.06	0.16	0.05	0.05
	2,3,4-trimethylpentane	0.03	0.92	0.13	0.05	0.88	0.09	0.04
*	toluene	0.01	32.7	6.3	4.11	67.7	17.18	16.18
	2-methylheptane	0.03	1.46	0.31	0.21	2.89	0.75	0.66
	3-methylheptane	0.02	0.57	0.1	0.07	1.25	0.26	0.20
	4-methylheptane, 1-methyl-1-cyclohexene	0.06	1.57	0.3	0.19	2.88	0.70	0.67
	1-octene	0.03	0.8	0.18	0.14	1.48	0.48	0.41
*	n-octane	0.02	1.82	0.34	0.23	2.97	0.87	0.75
*	ethylbenzene	0.02	6.06	1.38	1.14	11.1	3.26	3.14
	m-& p-xylene	0.04	22.2	5.14	4.26	40.4	12.1	11.5
*	styrene	0.03	1.02	0.22	0.16	3.83	0.60	0.54
*	o-xylene	0.04	7.72	1.81	1.5	14.8	4.59	4.42
*	α-pinene	0.07	1.95	0.41	0.27	2.30	0.88	0.78
*	n-propylbenzene	0.02	1.68	0.36	0.26	2.49	0.89	0.89
*	4-ethyltoluene	0.01	1.4	0.3	0.21	19.4	1.10	0.77
*	1,3,5-trimethylbenzene	0.03	2.79	0.58	0.38	23.3	1.83	1.41
*	n-decane	0.02	2.94	0.6	0.42	4.32	1.48	1.19
	1,2,4-trimethylbenzene, tet. butylbenzene	0.04	6.91	1.42	0.92	10.6	3.62	3.46
*	1,2,3-trimethylbenzene	0.01	1.31	0.27	0.17	2.09	0.72	0.63
	1,2,3,4-tetramethylbenzene	0.02	5.55	0.43	0.22	3.68	0.95	0.85

* Compounds included in the MCM model (see sec. 3.5).

As a result of seasonal change, the maximum $j(NO_2)$ and $j(O^1D)$ during winter are ~26 % and ~75 %, respectively lower than that during summer (see Fig. 3.1.4). The summer-to-winter ratio of maximum $j(O^1D)$ is comparable to that of ~3 for Tokyo (Kanaya et al., 2007) and ~4 for New York City (Ren et al., 2006). In addition, one hour after solar noontime $j(O^1D)$ and $j(NO_2)$ decreased by 25 % and 20 % of its maximum values, respectively during winter while they decreased by only 3 % and 4 %, respectively for the same time interval in summer.

Average mixing ratios of HONO, CO, NO and NO_2 were higher during winter than their corresponding values during summer (see Fig. 3.1.4). However, average and maximum mixing ratios of the photo-oxidation products O_3, PAN and HCHO (partially photochemically formed) during the summer were higher than their corresponding values during winter owing to the higher summertime photochemical activity (Fig. 3.1.4). The higher mixing ratios of the emitted species (CO, NO, NO_2) observed during the winter are due to the lower boundary layer height during winter (Gramsch et al., 2006). Jorquera, et al. (2002a) used a box model based on an air quality data base of Santiago to investigate CO, NO_x and CO_2 and showed that there is a strong reasonability in the meteorological factors that modulate air quality levels in the city of Santiago. In addition, during fall and winter seasons, the dispersion capacity of the city's air shed was found to be only 20-50 % of the summertime values. This typical seasonal dependency of the diurnal profiles of the emitted and photochemically formed species has previously been observed in New York City (Ren et al., 2006) and Tokyo (Kanaya et al., 2007).

Similar to summer, the $HONO/NO_x$ ratio shows two maxima. While the night-time maximum (~4:00 h) is typical for urban conditions and can be explained by night-time sources and the missing photolysis of HONO, the second maximum during daytime (14:00 h, see Fig. 3.1.4) points toward a strong daytime source of HONO, for which photochemical sources have been proposed (Kleffmann, 2007). This is also confirmed by the lower daytime $HONO/NO_x$ ratio during winter (lower photochemical activity) in comparison to that during summer.

The NMHCs profiles measured during winter were similar to those measured during summer (see sec. 3.1.1). The average diurnal profiles of NO and the total quantified NMHCs during the winter campaign are shown in Fig. 3.1.5.

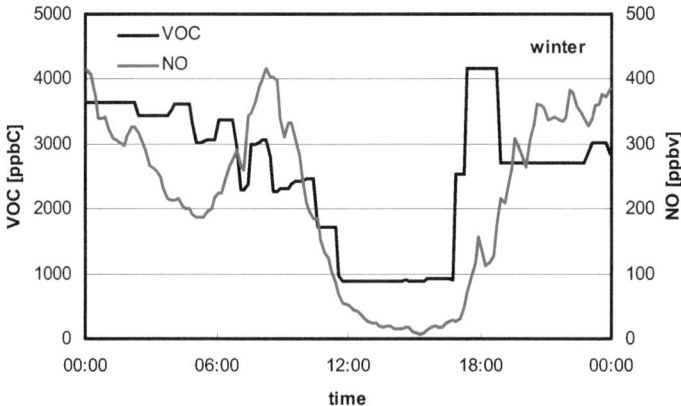

Fig. 3.1.5: Average diurnal variation of the total quantified VOCs and NO during the winter measurement campaign, 2005.

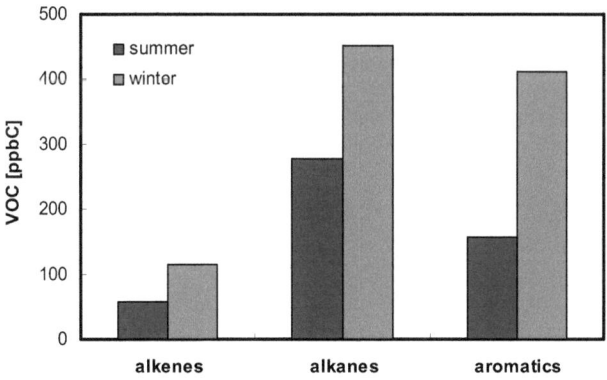

Fig. 3.1.6: Mixing ratios of the main categories of the measured VOCs during summer and winter in Santiago de Chile.

The identified NMHCs accounts on average for 48 % of the total quantified winter NMHCs of ~2000 ppbC in winter, which is >2 times higher than that of ~900 ppbC

Measurements results analysis

determined during the summer (see Fig. 3.1.6). Alkanes made the highest contribution during summer and winter of 56 % and 46 % followed by aromatics (32 % and 42 %) and alkenes (12 % and 12 %), respectively (see Fig. 3.1.7). The BTEX compounds (benzene, toluene, ethylbenzene and xylenes) contribute 80 % and 79 % to the total aromatics and 25 % and 33 % to the total identified NMHCs during summer and winter, respectively.

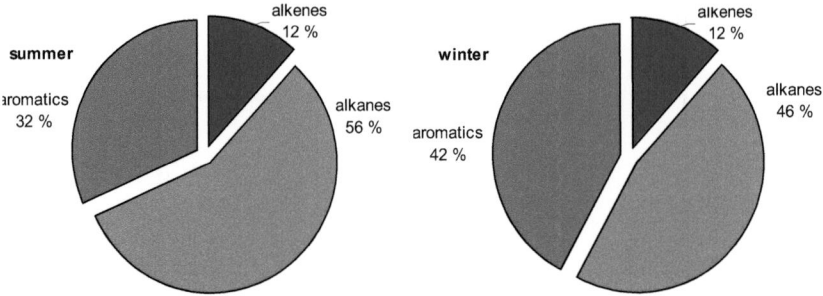

Fig. 3.1.7: Contributions of the different categories of hydrocarbons during the summer and winter in Santiago, 2005.

3.1.3 Emission Indices and Emission Ratios

As aforementioned (see introduction), emission indices are important parameters, which give information about the status of the air quality and traffic emissions. In the present study, emission indices for NMHCs and other trace gases were determined for the first time in Santiago during the winter campaign, 2005. For this purpose, NMHCs were measured at short time intervals ranging from 30 to 60 min during the rush hour time, for which significant variations of the concentrations of the trace gases of interest have been observed. These measurements were also accompanied with simultaneous measurements of CO_2, CO and NO_x (see Fig. 3.1.3).

In order to determine the emission indices (EI), the measured rush-hour concentrations were first corrected for background (i.e., contributions from other sources rather than that considered). The background was calculated employing a two-point linear regression equation for the measured rush-hour peak baseline. The emission value ($\Delta[X]$ or $\Delta[CO_2]$) was then calculated by subtracting the linear regression equation of the baseline from the measured peak concentrations.

E 3.1   emission value = measured peak concentrations – (baseline)

The emission ratios ($\Delta[X]/\Delta[CO_2]$) were calculated from the weigthed average of the slopes from the plots of ΔX against ΔCO_2 and are listed in Table 3.1.2. The $\Delta CO/\Delta CO_2$ value of 0.0187 ± 0.0027 (see Table 3.1.2) is comparable to that determined recently in Santiago (Rubio et al., 2009b). Comparably, the slopes of the correlation between all the measured values of CO and CO_2 (Fig. 3.1.8) and between all the emission values ΔCO and ΔCO_2 (see Fig. 3.1.9) are also shown. Both slopes are similar to that calculated using the method described above. However, this was not the case for all compounds, therefore; only the above method (i.e, weighted average of the slopes from the plot of ΔX against ΔCO_2) was considered.

The EI of the measured NMHCs and other pollutants (X) were calculated from the emission ratio ($\Delta[X]/\Delta[CO_2]$) and the molar masses of X and CO_2 using an emission index of 3138 g CO_2 kg^{-1} (Kurtenbach et al., 2001). The calculated EI are listed in Table 3.1.2.

High emission index of 1.1 g/kg was calculated for propane, which is mainly due to evaporative emissions from liquid petroleum gas (LPG) leakage (Chen et al., 2001). EIs for benzene, toluene, the sum of m-and p-xylene, and o-xylene were 0.30, 0.72, 0.83 and 0.32 g/kg fuel, respectively. EIs for CO, NO, NO_2, HONO and HCHO were 37, 4.72, 0.56, 0.047, 0.047 g/kg fuel (see Table 3.1.2). The EI of NO and NO_2 are in good agreement (within the uncertainty limits) with that previously reported by Kurtenbach et al. (2001). The value for HONO is ca. 2 times smaller than that of 0.088 ± 18 g/kg fuel reported by Kurtenbach et al. (2001) based on tunnel measurements. One possible reason for this difference could be the different emission sources in Santiago de Chile in comparison to Wuppertal, Germany due to the different vehicle/engines used. Diesel engines have typically higher EI for HONO than gasoline engines (Kurtenbach et al., 2001). However, since the number and type of the vehicle fleet was not investigated during the current study, this reason can not be verified.

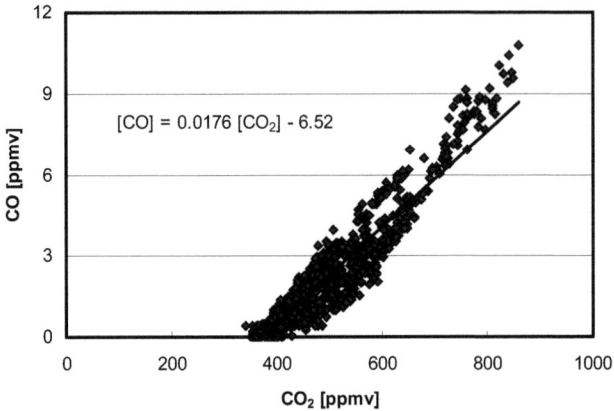

Fig. 3.1.8: Correlation between all the measured mixing ratios of CO and CO_2 in Santiago.

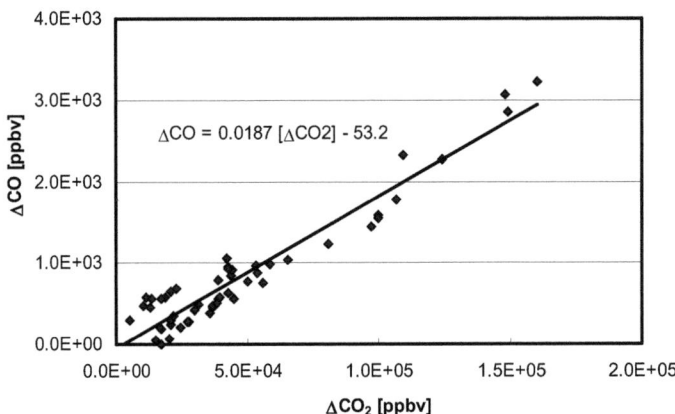

Fig. 3.1.9: Correlation between the emission values of CO (ΔCO) and CO_2 (ΔCO_2) during the morning rush hour periods.

Different NMHCs may have different life time due to their different reactivities towards the main oxidizing species in the atmosphere. The ratios between these hydrocarbons may thus reflect not only their emitting sources but also their photochemical environment (Monod et al., 2001). The correlation between benzene and toluene has been reported to be well through different sources as traffic, fuel samples, biomass burning but the slopes varied widely (Monod et al., 2001). During the winter campaign in Santiago, an average campaign toluene/benzene ratio of 2.6 (R^2, 0.95) was obtained, which reflects typical urban traffic fingerprints (Monod et al., 2001). Fig. 3.1.10 shows the good correlation between toluene and benzene demonstrating that their origin is combustion as it is the major source of toluene. This ratio is also in excellent agreement with that of 3.0 reported by Rappenglück et al. (2000).

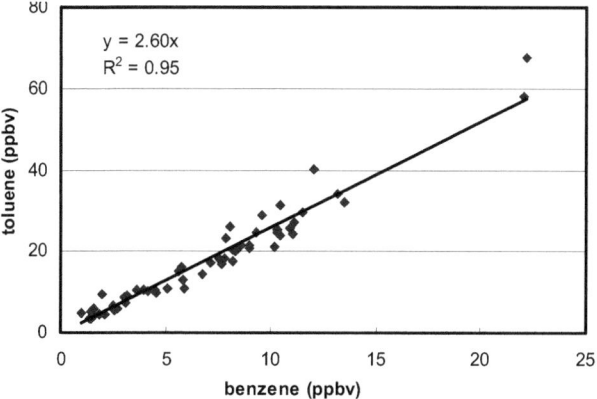

Fig. 3.1.10:   Correlation between average campaign toluene and benzene mixing ratios.

Table 3.1.2: Emission indices for the measured trace gases in the city of Santiago de Chile.

compound name	$\Delta X/\Delta CO_2$	SD [±]	EI (g/Kg Fuel)	+/-
propene	3.49E-05	2.05E-05	**0.105**	**0.062**
propane	3.62E-04	1.90E-04	**1.14**	**0.60**
i-butane	6.55E-05	3.88E-05	**0.272**	**0.161**
butadiene	3.98E-06	1.74E-06	**0.015**	**0.007**
n-butane	7.14E-05	7.91E-05	**0.296**	**0.328**
trans-2-Butene	1.82E-06	5.61E-07	**0.007**	**0.002**
cis-2-butene	2.34E-06	1.30E-06	**0.009**	**0.005**
3-methyl-1-butene	4.33E-06	3.32E-06	**0.022**	**0.017**
i-pentane	6.41E-05	4.15E-06	**0.330**	**0.021**
1-pentene	5.49E-06	1.27E-06	**0.027**	**0.006**
isoprene	4.05E-06	2.19E-06	**0.020**	**0.011**
trans-2-pentene	4.16E-06	2.61E-06	**0.021**	**0.013**
cis-2-pentene	3.25E-06	9.10E-07	**0.016**	**0.005**
2-methyl-2-butene	4.14E-06	1.84E-06	**0.023**	**0.010**
2,2-dimethylbutane	3.09E-05	8.74E-06	**0.190**	**0.054**
cyclopentene	6.86E-07	3.60E-07	**0.003**	**0.002**
2-methylpentane	6.21E-05	1.91E-05	**0.381**	**0.117**
3-methylpentane	3.49E-05	1.16E-05	**0.215**	**0.071**
1-hexene	4.65E-06	8.51E-07	**0.028**	**0.005**
2,3-dimethyl-1,3-butadiene	6.57E-07	2.25E-07	**0.004**	**0.001**
2,3-dimethyl-2-butene	2.51E-06	1.73E-06	**0.015**	**0.010**
benzene	5.43E-05	1.68E-05	**0.302**	**0.094**
2-methylhexane	8.63E-06	5.79E-06	**0.062**	**0.041**
cyclohexene	3.86E-06	1.28E-06	**0.023**	**0.008**
1-heptene	3.37E-06	7.39E-07	**0.024**	**0.005**
2,2,4-trimethylpentane	2.22E-05	1.64E-05	**0.181**	**0.134**
n-heptane	8.56E-06	2.66E-06	**0.061**	**0.019**
1,4-cyclohexadiene	1.07E-06	8.77E-07	**0.006**	**0.005**
2,3,4-trimethylpentane	5.93E-07	2.29E-07	**0.005**	**0.002**
toluene	1.09E-04	2.76E-05	**0.716**	**0.181**
2-methylheptane	9.01E-06	2.06E-06	**0.073**	**0.017**
3-methylheptane	1.59E-06	4.41E-07	**0.013**	**0.004**
1-octene	3.24E-06	1.57E-06	**0.026**	**0.013**
n-octane	5.45E-06	1.08E-06	**0.044**	**0.009**
ethylbenzene	2.79E-05	1.56E-05	**0.211**	**0.118**
m-&p-xylene	1.09E-04	5.86E-05	**0.828**	**0.444**
styrene	5.31E-06	3.35E-06	**0.039**	**0.025**
o-xylene	4.22E-05	2.97E-05	**0.319**	**0.225**
α-pinene	1.31E-05	6.33E-06	**0.127**	**0.061**
n-propylbenzene	4.31E-06	1.89E-06	**0.037**	**0.016**
4-ethyltoluene	7.02E-06	2.04E-06	**0.060**	**0.017**
1,3,5-trimethylbenzene	8.71E-06	8.86E-07	**0.075**	**0.008**
n-decane	1.11E-05	5.29E-06	**0.113**	**0.054**
1,2,3-trimethylbenzene	5.87E-06	1.40E-06	**0.050**	**0.012**
1,2,3,4-tetramethylbenzene	1.21E-05	7.40E-06	**0.115**	**0.071**
CO	1.87E-02	2.68E-03	**37.3**	**5.3**
HONO	1.39E-05	3.29E-06	**0.047**	**0.011**
HCHO	2.22E-05	1.93E-05	**0.047**	**0.041**
NO	2.21E-03	1.53E-03	**4.72**	**3.27**
NO_2	1.72E-04	8.01E-05	**0.564**	**0.263**

* EI calculated using an emission index of 3138 g CO_2 kg^{-1} (Kurtenbach et al., 2001).

The ratio of m, p-xylene/ethylbenzene (x/e) has been often used to determine the extent of the photochemical reactivity in an ambient air sample (Monod et al., 2001 and references therein). The investigation of the x/e ratio from different sources and from different locations has proven that they are emitted from the same sources namely, combustions and fuel evaporation. Because of their different reactivities toward OH radical and consequently their different atmospheric lifetimes, the x/e ratio is also a tool to investigate the photochemical age of an urban plume (Monod et al., 2001). An average ratio of 3.7 (R^2 = 0.99) was determined for m,p-xylene/ethylbenzene during the winter field campaign. This value is typical for direct urban traffic emissions while the excellent correlation (see Fig. 3.1.11) is an evidence of the same sources origin. This ratio is also in good agreement with that obtained in 2000 (Rappenglück et al., 2005).

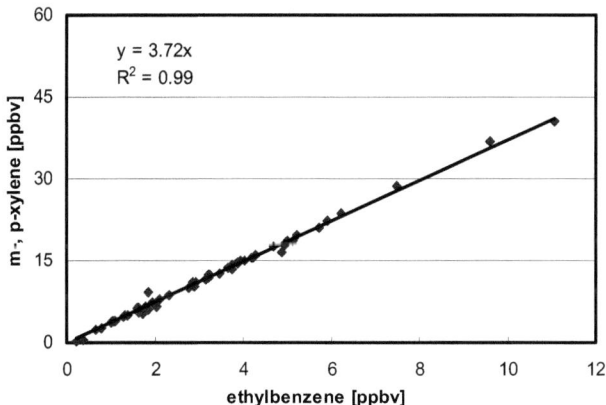

Fig. 3.1.11: Correlation between average campaign toluene and benzene mixing ratios.

The good correlation (R^2 = 0.96) between toluene and ethylbenzene (see Fig. 3.1.12) show that combustion emissions are the dominant common sources rather than solvent emissions.

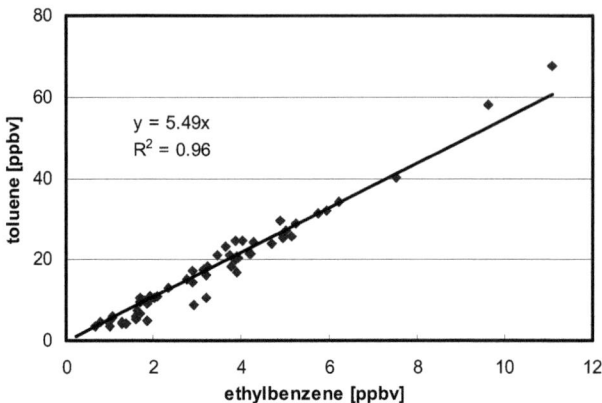

Fig. 3.1.12: Correlation between average campaign toluene and benzene mixing ratios.

The high contribution of alkanes to the NMHCs is driven by the high propane mixing ratio during both, summer and winter in Santiago (see Table 3.1.1). Propane median mixing ratio as high as 136 ppbv was also reported by Chen et al. (2001) for a measurement campaign in Santiago in June 1996. This high mixing ratio of propane has been reported to be a typical fingerprint of the liquefied petroleum gas (LPG) when coincided with high mixing ratios of n-butane and i-butane (Chen at al., 2001; Rappenglück et al., 2005), similar to the current study (see Table 3.1.1). This result shows that LPG is still being used intensively in Santiago. In general, it is concluded that the main NMHCs emission sources in Santiago are from traffics sources followed by LPG leakage and then fuel evaporation sources. This result agrees also with the receptor model evaluation done in Santiago in 1996 (Jorquera and Rappenglück, 2004).

3.1.4 Conclusion

In this section, the diurnal concentrations of measured species during two summer and winter measurement campaigns in Santiago de Chile were discussed. In addition, emission indices and emission ratios were determined for most of the measured species.

Average mixing ratios of HONO, CO, NO and NO_2 in winter were higher than their corresponding values during summer. However, average and maximum mixing ratios of the photo-oxidation products O_3, PAN and HCHO (partially photochemically formed) during summer were higher than their corresponding values during winter owing to the higher summertime photochemical activity. The higher mixing ratios of the emitted species (CO, NO, NO_2) observed during the winter are due to the lower wind speed and boundary layer height during winter.

During both, summer and winter campaigns, daytime HONO concentrations were significantly higher than in other polluted urban areas with the average HONO mixing ratio in winter higher than during summer. The high mixing ratios of HONO and the daytime maximum of the HONO/NO_x ratio in Santiago point to a very strong daytime HONO source. This is also confirmed by the lower daytime HONO/NO_x ratio during winter (lower photochemical activity) in comparison to that during summer.

As a result of seasonal change, the maximum $j(NO_2)$ and $j(O^1D)$ during winter are ~26 % and ~75 %, respectively lower than that during summer.

During summer and winter, identified NMHCs accounts on average for 48 % of the total quantified NMHCs of ~2000 ppbC in winter, which is >2 times higher than that of ~900 ppbC determined during the summer. Alkanes made the highest contribution of 46 % and 56 % during winter and summer followed by aromatics (42 % and 32 %) and alkenes (12 % and 12 %), respectively. The BTEX compounds (benzene, toluene, ethylbenzene and xylenes) contribute 79 % and 80 % to the total aromatics and 33 % and 25 % to the total identified NMHCs during winter and summer, respectively. The high contribution of alkanes to the NMHCs is driven by the high propane mixing ratio during both, summer and winter in Santiago. This high mixing ratio of propane has been reported to be a typical fingerprint of the liquefied petroleum gas (LPG) when coincided with high mixing ratios of n-butane and i-butane (Chen et al., 2001).

Emission indices for NMHCs and other trace gasses were determined for the first time in Santiago during the winter campaign, 2005. A high emission index of 1.10 g/kg was calculated for propane, which is mainly due to evaporative emissions from liquid petroleum gas (LPG) leakage. EIs for benzene, toluene, the sum of m- and p-xylene, and o-xylene were 0.17, 0.34, 0.34 and 0.13 g/kg fuel, respectively. EIs for CO, NO, NO_2, HONO and HCHO were 59, 6.93, 0.539, 0.044, 0.063 g/kg fuel. The emission index for the total quantified NMHCs in Santiago is 13.7 g/Kg fuel.

Emission ratios between the key VOC species were also determined in order to determine their sources and the extent of the photochemical reactivity in ambient air samples. The toluene/benzene ratio of 2.6 (R^2, 0.95) was obtained which reflects typical urban traffic fingerprints. An average ratio of 3.7 (R^2, 0.99) was determined for m-, p-xylene/ethylbenzene during the winter field campaign, which is typical for a direct urban traffic plume while the excellent correlation is an evidence of the same sources origin. The good correlation (R^2 = 0.96) between toluene and ethylbenzene show that combustion emissions are the dominant common source rather than solvent emissions. Therefore, the main NMHCs emissions sources in Santiago are mainly from direct traffic sources followed by LPG leakage and then fuel evaporation sources.

Generally, a strong seasonal dependence has been observed for both, photochemically formed products (e.g., O_3 and PAN) and primary emitted species (e.g., CO, CO_2, NO and VOCs). The effect of these seasonal changes on the photochemical oxidation process and radical budgets is discussed in details in the next section. The photochemical formation of ozone and its impacts on the air quality in Santiago is discussed in details in sec.3.3.

3.2 Oxidation Capacity and its Seasonal Dependence

In this section, a detailed analysis of the radical budgets, primary sources and sinks of the radicals in the highly polluted urban area of Santiago de Chile during the summer and winter measurement campaigns is presented. In addition, an investigation of the balance between secondary production and destruction of radicals for this study as well as for other urban studies is presented.

3.2.1 Oxidation Capacity

The oxidation capacity (OC) defined earlier (see introduction, sec. 1.3) has been calculated from the loss rate of the VOCs and CO due to reactions with OH, O_3 and NO_3, which were estimated using the MCMv3.1 box model (see 2.3.2).

During summer, the average oxidation capacity of OH, O_3 and NO_3 radicals through the entire day represented 89.4, 10.3 and 0.3 % of the total oxidation capacity, respectively. Clearly, the OH radical was the dominant oxidant during daytime contributing by a maximum of 3.2×10^8 molecule cm^{-3} s^{-1} (94 %) to the total oxidation capacity at about 15 h. The ozone contribution to the oxidation capacity during daytime ranged from 6 % to 11 % while it reached >50 % during the night, mainly due to alkene ozonolysis. In general, the nitrate radical had a negligible contribution during both, day and night, which was mainly caused by the high NO concentrations during the campaign. The total number of the depleted molecules per day due to oxidation by OH, O_3 and NO_3 were 6.4×10^{12}, 7.4×10^{11} and 2.0×10^{10} molecules cm^{-3} during summer and 3.6×10^{12}, 2.3×10^{11} and 2.7×10^{10}, during winter, respectively.

During winter, the average oxidation capacity of OH, O_3 and NO_3 radicals throughout the entire day represented 93.4, 5.9 and 0.7 % of the total oxidation capacity, respectively. During daytime, maximum OH, O_3 and NO_3 oxidation rates of 1.4×10^8, 2.5×10^7 and 2.8×10^6 molecules cm^{-3} s^{-1} were reached at 13:00 h, 17:00 h and 15:00 h, respectively. During the night, OH is also the dominant oxidant with an average contribution of 87 % followed by O_3, ~12 %, while NO_3 had negligible contributions to the total OC. These results are in contrast to those calculated under summertime conditions where O_3 had a contribution of up to >50 % during the night. The main reason for this difference is the extremely high NO_x concentration during winter in comparison to that during summer, which more efficiently suppress night time levels of O_3 and NO_3 through the reactions $O_3 + NO \rightarrow NO_2$ and $NO_3 + NO \rightarrow 2NO_2$.

The total number of the depleted molecules per day due to oxidation by OH, O_3 and NO_3 were 3.6×10^{12}, 2.3×10^{11} and 2.7×10^{10} molecules cm^{-3}, respectively, which is much lower than those reported during summer. Owing to the dominant contribution of the OH radical to the OC, the current study focuses only on the analysis of production and destruction rates of the OH radical.

3.2.2 Radical Production and Destruction Rates

The total production and destruction rates of OH and HO_2 calculated by the MCM model during summer and winter are shown in Fig. 3.2.1 and Fig. 3.2.2 in addition to the ratios of the radical production/destruction. The ratio was around unity throughout the day for both, the OH and HO_2 radicals. The total production and destruction rates of both radicals in summer are ~2 times higher than those during winter. Reasons for this difference are mainly due to the lower photochemical activity and the much higher NO_x levels observed during winter (see also sec. 3.1.2). NO_2 is the main dominant sink of OH radicals in Santiago as well as in other urban areas and thus, higher NO_2 levels reduce the OH radical concentration in winter. Emmerson et al. (2005b) reported that termination of OH in Birmingham city centre, UK was almost entirely dominated by nitrogen species with a factor of 2 higher in winter than in summer. Besides this, Emmerson et al. (2005a) have shown that doubling the NO_2 concentrations on polluted days has lead to a decrease in the modelled OH and HO_2 concentrations by ~65 % and ~62 %, respectively. Recently, Dusanter et al. (2009) also showed that decreasing and increasing the NO_x concentrations by a factor of 2 changed the predicted OH and HO_2 levels in the range +44/-43 % and +157/-68 %, respectively.

The diurnal profiles of the total production and destruction rates of both OH and HO_2 in winter are slightly different than those during summer and are characterized by a second afternoon peak at ~17:00 h (see Fig. 3.2.2) which is mainly due to the elevated level of pollutants during winter. The high VOC levels in the presence of high NO concentrations (see sec 3.1.2) that is overlapped by a decreasing actinic flux leads to the sharp afternoon peaks in the HO_2 and OH propagation rates owing to the efficient recycling of the peroxy radicals ($RO_2 + HO_2$) with NO.

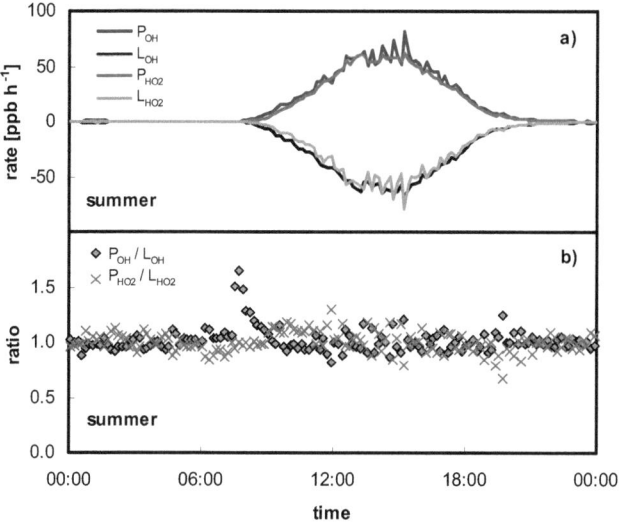

Fig. 3.2.1: a) Production and destruction rates of OH and HO_2 in summer. b) Ratios of production to destruction rate of OH and HO_2 during summer.

During both, summer and winter, total radical production and destruction rates were dominated by the recycling reactions of peroxy radicals (RO_2 + NO and HO_2 + NO) and the oxidation of hydrocarbon, L_{OH}(OH+VOC), respectively (see Fig. 3.2.3 and Fig. 3.2.4). The main RO_2 production term is due to hydrocarbon oxidation with OH (hereafter referred as L_{OH}(OH→RO_2) with an average rate of 24.9 ppbv h^{-1} during summer in comparison to 12.6 ppbv h^{-1} during winter. The main loss of RO_2 during summer and winter is due to its reaction with NO (L_{RO2}(RO_2+NO)) with an average daytime loss rate of ~34.6 ppbv h^{-1} and ~15.1 ppbv h^{-1}, respectively which accounts for most of the HO_2 production. Other important HO_2 sources (hereafter totally referred as L_{OH}(OH→HO_2) are the reactions of OH with oxygenated VOCs (OVOCs except HCHO), CO and HCHO with average daytime production rates of 1.95, 0.48 and 1.1 ppbv h^{-1} during summer and 0.83, 0.79 and 0.31 ppbv h^{-1} during winter. In contrast to other secondary oxygenated VOCs (OVOCs), photolysis of HCHO has been considered as a net source of HO_2 (P_{HO2}(prim), see sec. 2.3.1) with summer and winter average daytime production rate of 0.54 and 0.23 ppbv h^{-1}, respectively (see Fig. 3.2.5).

Oxidation capacity and its seasonal dependence

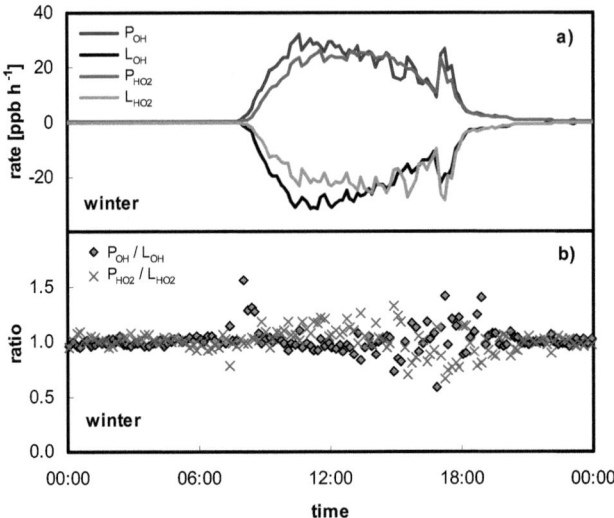

Fig. 3.2.2: a) Production and destruction rates of OH and HO$_2$ in winter. b) Ratios of production to destruction rate of OH and HO$_2$ during winter.

The main destruction route of HO$_2$ is through its reaction with NO (L$_{HO_2}$(HO$_2$+NO)) with a daytime average of 28.4 and 14.1 ppbv h^{-1}, which corresponds to 90 % and 79 % of the total destruction rate during summer and winter, respectively. These relative values are comparable to the TORCH summer campaigns of ~99 % (Emmerson et al., 2007) and BERLIOZ, >80 % (Mihelcic et al., 2003). The next most important HO$_2$ destruction path is due to its reaction with NO$_2$ to form HO$_2$NO$_2$ with an average daytime loss rate of 3.2 and 3.3 ppbv h^{-1} during summer and winter, respectively. This loss path however, is essentially a reversible reaction that leads to HO$_2$ formation with an average daytime rate of 3.3 and 3.4 ppbv h^{-1} during summer and winter, respectively. The loss rates due to the HO$_2$ self-reaction and its cross-reactions with RO$_2$ are very negligible during both, summer and winter, in agreement with other urban studies (e.g., George et al., 1999; Ren et al., 2006).

The main OH loss route is through its reaction with hydrocarbons, followed by reactions with NO and NO$_2$. The rates of OH destruction due to hydrocarbons oxidation depend on the detailed chemical mechanism and can be estimated using the following relationships:

E 3.2 $L_{OH}(OH+VOC) \approx L_{OH}(total) - k_{NO_2+OH}[NO_2][OH] - k_{OH+NO}[NO][OH]$

E 3.3 $L_{OH}(OH+VOC) \approx \sum k_i [VOC_i][OH]$,

where L_{OH} (total) is the total loss rate as calculated by the MCM model and k_i represents the bimolecular rate constant for OH reaction with the corresponding VOC_i.

Alternatively, since OH loss due to the oxidation of CO and hydrocarbons is dominated by the fluxes of $L_{OH}(OH \rightarrow HO_2)$ and $L_{OH}(OH \rightarrow RO_2)$, $L_{OH}(OH+VOC)$ can be also calculated in terms of these radical fluxes:

E 3.4 $L_{OH}(OH+VOC) \approx L_{OH}(OH \rightarrow HO_2) + L_{OH}(OH \rightarrow RO_2)$

$L_{OH}(OH \rightarrow RO)$ is not considered here, because of its negligible contribution (Emmerson et al., 2005b, 2007).

If equation E 3.3 is used to calculate $L_{OH}(OH+VOC)$ due to reactions with the measured VOCs only, the OH loss rate will be underestimated since reactions with secondary VOC products are not included. Consequently, HO_2 as a net source of OH will be overestimated. Equations E 3.2 and E 3.4 takes into account the detailed degradation of the VOCs due to reactions with OH, as calculated by the MCM photochemical box model, which includes the secondary VOC oxidation products. The summer average daytime loss of OH radicals by VOC reaction calculated employing equation E 3.3 is about 6 ppbv h^{-1} while that obtained using equation E 3.2 and E 3.4 is 28.4 ppbv h^{-1}, representing about 79 % of the total OH loss. The winter average daytime loss rate of OH radicals by VOC reaction calculated employing the equations E 3.2 and E 3.4 is ~14.5 ppbv h^{-1}, representing 67 % of the total OH loss, a relative contribution which is lower than in summer. The fraction of OH loss by VOC reactions is similar to that calculated for Berlin, 50-70 % (Mihelcic et al., 2003) and Mexico City, 72 % (Shirley et al., 2006).

OH production is dominated by the recycling reaction of HO_2 with NO, $P_{OH}(HO_2 \rightarrow OH)_{recycled}$ for which:

E 3.5 $P_{OH}(HO_2 \rightarrow OH)_{recycled} = P_{OH}(HO_2 \rightarrow OH) - P_{HO_2}(prim)$,

where:

E 3.6 $P_{OH}(HO_2 \rightarrow OH)$ = $k_{HO_2+NO}[HO_2][NO]$.

$P_{OH}(HO_2 \rightarrow OH)_{recycled}$ reached a maximum production rate of 70.5 and 21.4 ppbv h^{-1} with a daytime average of ~27.8 and ~13.9 ppbv h^{-1} during summer and winter, respectively (Fig. 3.2.5). The $P_{OH}(HO_2 \rightarrow OH)$ route accounts for 80 % of the total OH radical production during summer which is higher than that of ~66 % during winter. The summer value is comparable to that of TORCH, 80 % (Emmerson et al., 2007) and BERLIOZ, >70 % (Mihelcic et al., 2003) simulated during summer and that of Mexico City, >80 % (Shirley et al., 2006 and Sheehy et al., 2008) during spring. Production of OH due to HO$_2$ reaction with O$_3$ is negligible for both, summer and winter with ~0.014 ppbv h^{-1} and <0.01 ppbv h^{-1}, respectively. As aforementioned, the lower contribution of L$_{OH}$(OH+VOC) and P$_{OH}$(HO$_2 \rightarrow$OH) to the total OH loss and production rates, respectively, during winter compared to that during summer is due to the higher NO$_x$ levels during winter. High concentrations of NO$_2$ suppress the OH levels through the (NO$_2$+OH) reaction and consequently decrease the hydrocarbons oxidation rate, therefore decreasing P$_{OH}$(HO$_2 \rightarrow$OH). In addition, the lower HO$_2$ formation rate from photolytic sources (e.g., photolysis of carbonyls) as a result of the lower actinic flux in winter may also lead to a decrease in the flux of P$_{OH}$(HO$_2 \rightarrow$OH).

Oxidation of hydrocarbons results in the production of O$_3$ (through the NO catalytic recycling of RO$_2$ and HO$_2$) and HCHO (partial secondary oxidation product) as by-products in addition to alkene ozonolysis as a subsequent process. However, due to their lower photochemical production and photolysis frequencies during winter (see sec. 3.1.2), these processes have a lower contribution to OH formation than during summer (see sec. 3.2.6). These processes, in addition to HONO photolysis, constitute the net radical production term, P$_R$ (see Fig. 3.2.3, Fig. 3.2.4 and Fig. 3.2.5). The rest of the OH production term (P$_{OH}$ (others), 1.1 and 0.50 ppbv h^{-1} in summer and winter, respectively) is mainly due to secondary sources, e.g., photolysis of >1000 compounds of ROOH and RCO$_3$H in the model that result from RO$_2$ cross reactions with HO$_2$ in addition to some contribution from ozonolysis of secondary alkenes. However, these secondary sources are not constrained by the measurements and do not add to the net radical sources (see sec. 3.2.5). The balance between P$_{OH}$(HO$_2 \rightarrow$OH)$_{recycled}$ and L$_{OH}$(OH+VOC) (see sec. 3.2.4), results in the NO$_2$ + OH (termination) reaction becoming the net dominant sink for OH with a maximum loss rate of 6.4 and 7.1 ppbv h^{-1} and a daytime average loss rate of ~3.4 and ~3.6 ppbv h^{-1} during summer and winter, respectively (see Fig. 3.2.5). Ozone sensitivity analysis performed during the summer campaign (see sec. 3.3), showed that only under very low NO$_x$ conditions reaching <5 % of the current levels, HO$_2$ recycling through its reaction with NO could be a limiting factor. Under these conditions hydrocarbon oxidation could be a net sink for OH radicals, which in turn will also lead to a reduction in the OH sources, i.e. O$_3$ and HCHO photolysis as well as alkenes ozonolysis.

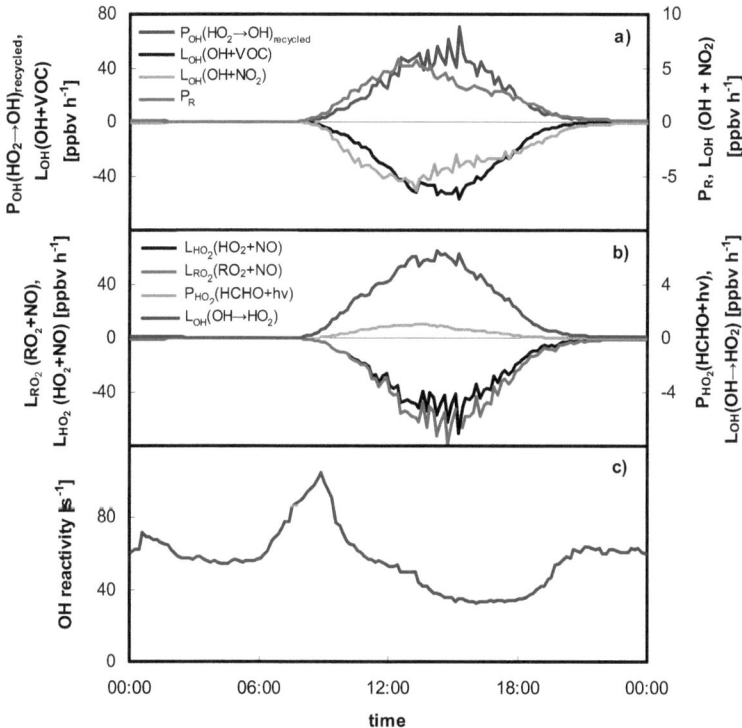

Fig. 3.2.3: Average day production and destruction rates of a) OH and b) HO$_2$ and c) Modelled OH reactivity during summer in Santiago.

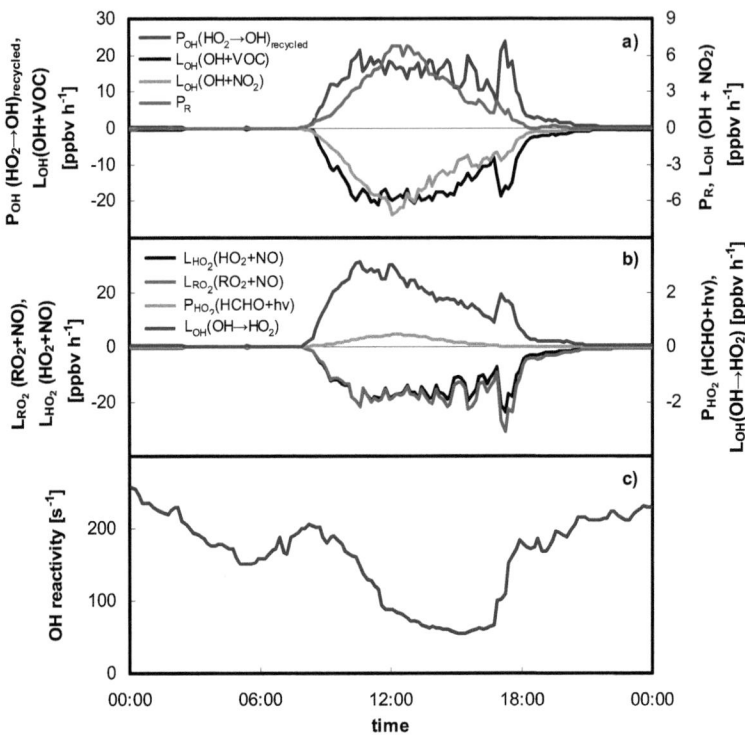

Fig. 3.2.4: Average day production and destruction rates of a) OH and b) HO$_2$ and c) Modelled OH reactivity during winter in Santiago.

3.2.3 OH Reactivity

The OH reactivity defined as the reciprocal of the OH radical lifetime has been calculated from the following relation:

E 3.7 $\quad$ OH reactivity $= \dfrac{L_{OH}\,(\text{total})}{[OH]}$

The diurnal profile of the OH reactivity in summer and winter (see Fig. 3.2.3 and Fig. 3.2.4) is characterized by two peaks corresponding to both, morning and night rush hour times which is quite similar to observations in New York City (Ren at al., 2006) and Mexico City (Shirley et al., 2006). The mean day average modelled OH reactivity during summer is about 42 s^{-1} reaching a maximum of 10^5 s^{-1} during rush hour with a nighttime peak of 60 s^{-1} (see

Fig. 3.2.3). The OH reactivity during summer is slightly higher than the average and night-time peaks measured during spring in Mexico City of 25 and 35 s^{-1}, respectively during spring, while the maximum measured OH reactivity in Mexico City of 120 s^{-1} during morning rush hour exceeded that of Santiago (Shirley et al., 2006). Sheehy et al. (2008) have also reported a modelled total reactivity of 110 s^{-1} during the morning rush hour and 45-50 s^{-1} at night in Mexico City.

During winter, the daytime average modelled OH reactivity of 90 s^{-1} (Fig. 3.2.4) is about two times higher than that during summer which is due to the apparent higher pollutant levels during winter in comparison to the summer (see sec. 3.1.2). Despite the lower absolute values in New York City compared to those in Santiago, the OH reactivity was also much higher in winter than in summer (Ren et al., 2006). The OH reactivity in winter reach a daytime maximum of ~200 s^{-1} during rush hour with a broad night-time peak that reach up to 255 s^{-1} at midnight (Fig. 3.2.4) due to the high level of pollutants at this time (see Fig. 3.1.4 and Fig. 3.1.5).

As a result of this high OH reactivity in winter, OH has a daytime average lifetime, τ_{OH} of ~0.01 s in comparison to ~0.02 s during summer which is lower than that of 0.04 s reported in the MCMA-2006 campaign (Dusanter et al., 2009, supplementary data). The daytime average lifetime of OH, τ_{OH} due to only the reaction with NO$_2$ during winter is 0.07 s in comparison to that of 0.24 s during summer caused by the much higher NO$_2$ levels during winter.

Underestimation of the OH reactivity using relationship, calculated based solely on the measured trace gas concentrations, has been previously observed when compared with the measured OH reactivity in different field measurements (Di Carlo et al., 2004; Yoshino et al., 2006; Ren et al., 2006 and references therein). It is worth mentioning that the OH uptake on aerosol surfaces and the uncertainty of the rate coefficient $k_{(NO_2 + OH)}$ could not account for the missing OH reactivity in previous field measurements (Yoshino et al., 2006).

3.2.4   Radical Propagation

Although hydrocarbon oxidation consumes most of the OH radicals (L_{OH}(OH+VOC) = 28.4, 14.5 ppbv h^{-1}), it also regenerates these radicals through the secondary production of OH, P_{OH}(sec.), (28.9, 14.4 ppbv h^{-1}) given by the sum of P_{OH}(HO$_2$→OH)$_{recycled}$ (27.8, 13.9 ppbv h^{-1}) and other secondary sources of 1.1, 0.5 ppbv h^{-1} on average during summer and winter, respectively (see Fig. 3.2.5). This balance shows that secondary production of OH (e.g. from HO$_2$ and RO$_2$ initiation sources) does not significantly add to the net OH initiation sources.

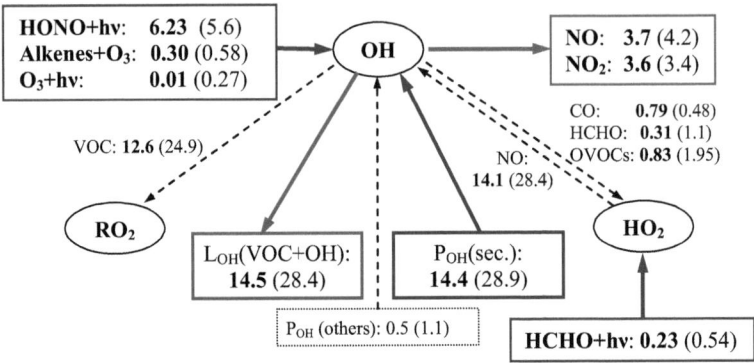

Fig. 3.2.5: Average daytime fluxes of the radical sources and sinks for the city of Santiago de Chile calculated by the MCM model during winter (bold) and summer (in between brackets). See text for definitions. Units are in ppbv h^{-1}. HONO net contribution, 82 % (52 %) in this diagram is slightly lower than in the text because [HONO]$_{PSS}$ is calculated here using [OH] calculated by the MCM (3 - 4 % higher).

To further investigate the recycling process, an additional MCM model scenario has been used, in which the concentrations of all aromatic hydrocarbons and alkanes in addition to isoprene and propene have been increased by a factor of 2 while the rest of the alkenes have been left unchanged. This additional sensitivity run represents more realistic conditions because not all quantified VOCs could be identified and not all defined hydrocarbons could be included in the MCM model because either some of these compounds were measured as a mixture of two or more compounds or were not defined in the MCM (see Table 3.1.1). The reason for including only isoprene and propene is because of their relatively high reactivity with OH but their low potential for OH production through ozonolysis (see sec. 3.2.9.2). Simulated OH only increased by ~1 % in both winter and summer for this additional modelling scenario. In addition, although both, HO$_2$ and RO$_2$ concentrations increased by ~80 % during summer and ~60 % during winter, the daytime average RO$_2$/HO$_2$ ratio remained unchanged during summer and winter. These results also demonstrate that the main net radical sources and sinks were not affected by the VOC level and that the potentially important secondary radical sources (e.g. OVOC photolysis) and sinks (e.g. $RONO_2$ formation) cancel each other, i.e. do not affect the OH radical concentration. In the main, this can be explained by the high NO concentrations during daytime in Santiago and the fast recycling through the reactions RO_2+NO and HO_2+NO. Similarly, Dusanter et al. (2009) reported that similar predicted HO_x concentrations were obtained when using a small or large dataset of VOCs keeping alkenes almost constant in both scenarios.

In addition, despite the different emission profiles during summer and winter (see sec. 3.1.2), the RO_2/HO_2 ratio was similar during both seasons ranging from 1-1.5 with an average daytime value of 1.3 and 1.2 during both, summer and winter, respectively. According to the MCM, the increase of the VOC levels under the high NO_x levels in Santiago during winter did not significantly affect the RO_2/HO_2 ratio. The maximum RO_2/HO_2 ratio (Fig. 3.2.6) is similar to that reported in Berlin with a maximum-modelled ratio of 1.3 (Mihelcic et al., 2003) but much lower than that of 3.9 calculated for the TORCH campaign (Emmerson et al., 2007).

The high recycling efficiency of the peroxy radicals during summer and winter can be demonstrated by the low average HO_2/OH ratio of 8 and 9 during summer and winter, respectively (see Fig. 3.2.6). The low value of the HO_2/OH ratio in summer and winter is typical for highly polluted conditions (e.g. Mihelcic et al., 2003) and implies a high recycling efficiency towards OH. While both, the RO_2/HO_2 and HO_2/OH ratios reach a minimum during the morning rush hour, only the HO_2/OH ratio reaches its maximum during early afternoon when the NO levels reach a minimum.

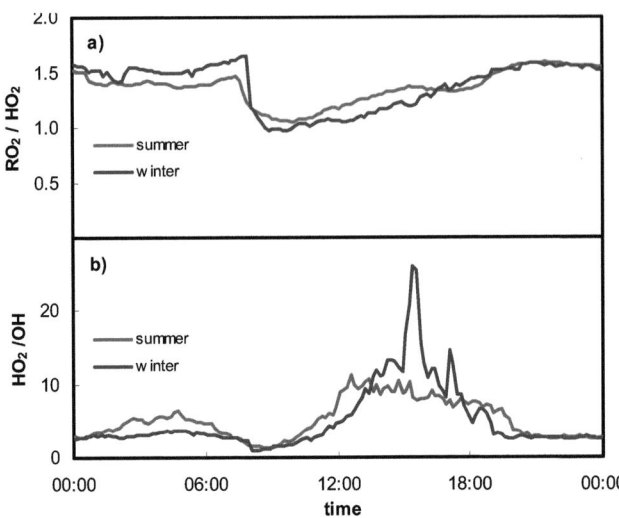

Fig. 3.2.6: Average diurnal variation of a) RO_2/HO_2 and b) HO_2/OH during summer and winter.

The average daytime maximum total peroxy radical concentration of 15 pptv and 5.3 pptv during summer and winter (see Fig. 3.2.7), respectively is relatively low when compared with other studies (Mihelcic et al., 2003; Shirley et al., 2006) and can be explained by the high NO concentrations in the city of Santiago. The maxima in the diurnal profiles of RO_2 and HO_2 during the winter and summer (Fig. 3.2.7) also coincide with the NO minima as shown in Fig. 3.1.4 (p. 58). This is also in agreement with the expected anti-correlation between the HO_2/OH ratio and NO as shown in Fig. 3.2.8, which is in accord with other studies (e.g. Emmerson et al., 2007; Dusanter et al., 2009 and references therein).

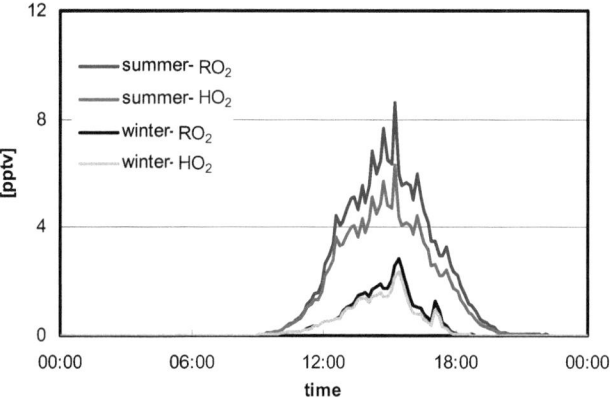

Fig. 3.2.7: Average diurnal profiles of RO_2 and HO_2 during summer and winter.

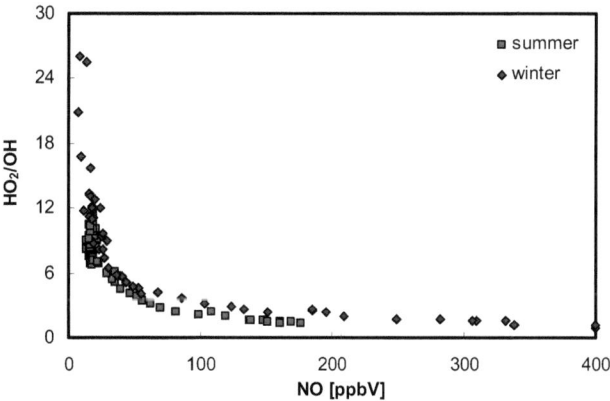

Fig. 3.2.8: Average daytime correlation between HO_2/OH and NO for summer and winter.

3.2.5 Balance Ratio and Comparison with other Studies

To further investigate the balance between secondary radical production and destruction for other studies, for which secondary radical fluxes are available, a balance ratio has been defined. This ratio is henceforth denoted as BR and calculated as follow:

$$\text{E 3.8} \quad BR = \frac{P_{OH}(HO_2 \to OH)}{L_{OH}(OH \to HO_2) + L_{OH}(OH \to RO_2)}$$

The secondary production term in the above equation is given by the $P_{OH}(HO_2 \rightarrow OH)$ instead of $P_{OH}(sec.)$ which was used in sec. 3.2.4. That is because $P_{OH}(others)$ (see sec. 3.2.2) was not reported in the studies under investigation (probably because of its low contribution to OH) and HCHO photolysis has a very small contribution in comparison to the total flux of $P_{OH}(HO_2 \rightarrow OH)$.

According to this balance ratio, a BR value of ~1 show that secondary production and destruction of OH are balanced. Therefore, RO_2 and HO_2 initiators should not be considered as net radical sources. That is because all the consumed OH during the oxidation process of the VOCs and CO has been reproduced through RO_2 and HO_2 recycling without a significant additional secondary OH production. A BR <1 shows that secondary radical production is smaller than destruction. This case may occur under very low NO_x conditions, which may lead to a low radical recycling efficiency and consequently, low secondary OH production. The secondary production of radicals is higher than their secondary destruction for BR >1. Accordingly, secondary radical initiators (e.g., photolysis of OVOCs) may be considered as net radical sources.

For this analysis, the data reported for summer (1999) and winter (2000) of PUMA field campaigns in Birmingham city centre, UK (Emmerson et al., 2005b), TORCH summer (2003) campaign at Writtle college, 25 miles north-east of London (Emmerson et al., 2007), the MCMA-2003 (2003) spring campaign in Mexico City (Sheehy et al., 2008) and the current study have been used. In all of the above studies, the MCM was used as the primary modelling tool. The slope of the correlation between average values of the secondary OH production given by $P_{OH}(HO_2 \rightarrow OH)$ and the secondary OH radical destruction, $L_{OH}(OH+VOC)$ given by $(L_{OH}(OH \rightarrow HO_2) + L_{OH}(OH \rightarrow RO_2)$ gives the average BR for the above studies as shown in Fig. 3.2.9.

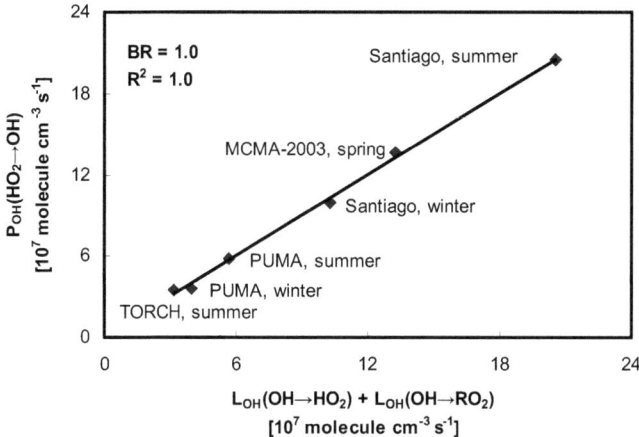

Fig. 3.2.9: Correlation between the average secondary radical production and destruction rates for different studies (for references see text).

Interestingly, an average BR value of 1.0 was obtained for the investigated studies with an excellent correlation coefficient (Fig. 3.2.9). For campaigns for which secondary fluxes were available for several days, average BR values of 1.03±0.01 (MCMA-2003), 1.09±0.02 (TORCH, summer), 0.99±0.02 (PUMA, summer), 0.90±0.02 (PUMA, winter) were obtained. For the average day campaigns in Santiago, BR values of 1.00 and 0.97 were obtained during summer and winter, respectively. This result shows that despite the widely different absolute values of radical production and destruction rates due to the different concentrations of constraints in the MCM models used under different atmospheric conditions, there appears to be on average a balance between the secondary OH production and destruction. Thus, the radical balance given by the BR value of 1.0 shows that initiation sources of HO_2 and RO_2 typically do not represent a net radical source which is in agreement with our previous conclusion (see sec. 3.2.4).

An error margin of the BR values of ±0.10 has been estimated from the standard deviation (σ) of the average BR value for the above campaigns at 95 % confidence limit (2.571×σ). Thus, for BR values in the range 0.9 - 1.1 no net radical sources or sinks can be identified within the estimated errors.

In order to determine the sensitivity of the BR values to additional secondary OH sources (e.g. photolysis of OVOCs), HO_2 initiation rates have been considered as net radical sources as has been proposed in other studies (Volkamer et al., 2007). Thus, adding the HO_2 initiation source to $P_{OH}(HO_2 \rightarrow OH)$, an increase in the BR value for PUMA-summer from 0.99 to 1.20, MCMA-2003-spring (from 1.03 to 1.18), Santiago-summer (from 1.0 to 1.19) and Santiago-winter (from 0.97 to 1.14) was obtained. In addition, since contributions from the RO_2 and RO initiation rates are not included, these values should be considered as a minimum limit for the contribution from secondary sources. Obviously, these hypothetical BR values are higher than the above estimated error margin of 1.10. Thus, our conclusion that the secondary HO_2 sources do not contribute to the net initiation sources of OH is also fulfilled for the other investigated studies. Similarly, Dusanter et al. (2009) showed that increasing ketone and aldehyde (higher than HCHO) concentrations by 200 % and 300 % lead to an increase in the OH concentration by only 0.6 % and 4.2 %, respectively, while HO_2 increased in the later by 16 % - 21 %.

These results demonstrate that the OH radical concentration is only determined by its primary sources and permanent sinks. In contrast, the concentrations of HO_2 and RO_2 are controlled by the VOC secondary chemistry. This is confirmed by the model simulations in which the VOC concentrations were almost doubled. In these simulations, the OH level was not affected during both, summer and winter while HO_2 and RO_2 increased by ~80 % during summer and ~60 % during winter (see sec. 3.2.4). This result may also explain the similar measured OH concentration observed at different HO_2 and RO_2 levels, which were obtained under different atmospheric conditions during the PUMA and TORCH campaigns (Emmerson et al., 2005a, 2007, respectively).

Fig. 3.2.9 also shows that the propagation rate during summer were generally much higher than those in winter for the same measurement site. In Santiago, the propagation rate during summer is ~2 times higher than in winter (see also sec. 3.2.4) while for the PUMA campaigns, it was ~1.6 times higher in summer than winter (Fig. 3.2.9). During winter and summer, Santiago has the highest propagation rates followed by MCMA-2003, PUMA and finally the TORCH campaign.

3.2.6 Net Radical Sources

As a result of the balance between secondary radical production and destruction, the total rates of radical initiation and termination can be calculated by considering only the net radical sources and sinks incorporated in the simple steady state approach (see sec. 2.3.1). The rates of the radical initiation, P_R, and termination, L_R, were calculated using the same rate constants incorporated in MCMv3.1.

The average absolute and relative diurnal contributions of the primary sources to radical production during summer and winter are shown in Fig. 3.2.10, Fig. 3.2.11, Fig. 3.2.12 and Fig. 3.2.13. For daytime conditions during summer and winter, net HONO photolysis, $P_{OH}(HONO) = j(HONO)[HONO] - k_{OH+NO}[NO][OH]$, has the highest average contribution of 55 % and 84 %, respectively followed by alkenes ozonolysis, $P_{OH}(alkenes)$ (24 %, 9 %), HCHO photolysis, $P_{OH}(HCHO)$ (16 %, 6.5 %) and ozone photolysis, $P_{OH}(O_3)$ (5 %, 0.5 %).

The high relative contribution of $P_{OH}(HONO)$ during summer and winter is in agreement with other recent studies (Ren et al., 2003, 2006; Kleffmann et al., 2005; Acker et al., 2006a, 2006b). For average daytime conditions, high net mean and maximum $P_{OH}(HONO)$ of 2.9 ppbv h^{-1} and 6.2 ppbv h^{-1} have been determined during winter, which is much higher than those of 1.7 ppbv h^{-1} and 3.1 ppbv h^{-1}, respectively, determined during summer. These values are even higher than the ~2 ppbv h^{-1} reported by Ren et al. (2003) for New York City during summer. Only in the study of Acker et al. (2006b) a similar high maximum OH production rate by HONO photolysis of up to 6 ppbv h^{-1} reported for the city of Rome. However, this number is an upper limit since in their estimations the back reaction of NO+OH was not considered. Especially during the morning, for which the maximum production rate was reported by Acker et al. (2006b), high NO concentrations can lead to a strong overestimation of net OH production rates (see sec. 3.2.9.3).

Oxidation capacity and its seasonal dependence

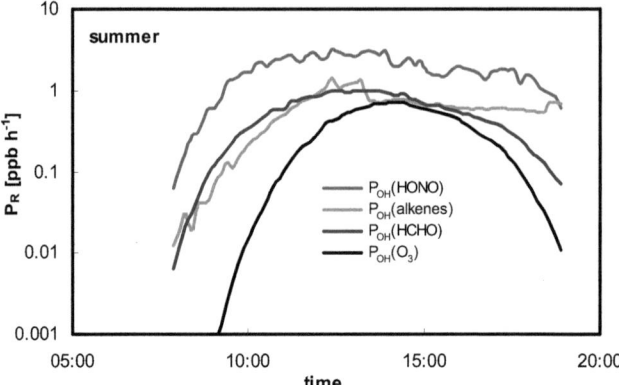

Fig. 3.2.10: Average absolute diurnal daytime radical production rates (P_R) during summer in Santiago.

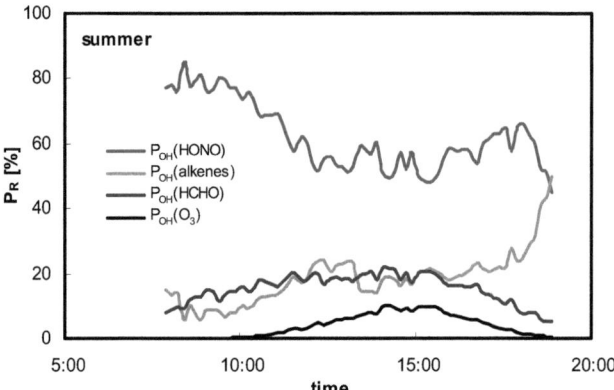

Fig. 3.2.11: Average relative diurnal daytime radical production rates (P_R) during summer in Santiago.

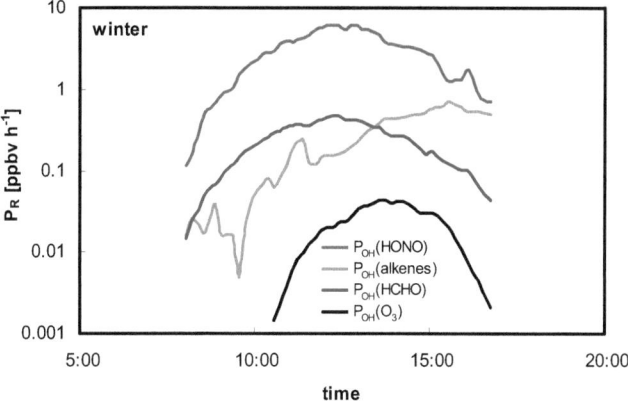

Fig. 3.2.12: Average absolute diurnal daytime radical production rates (P_R) during winter in Santiago.

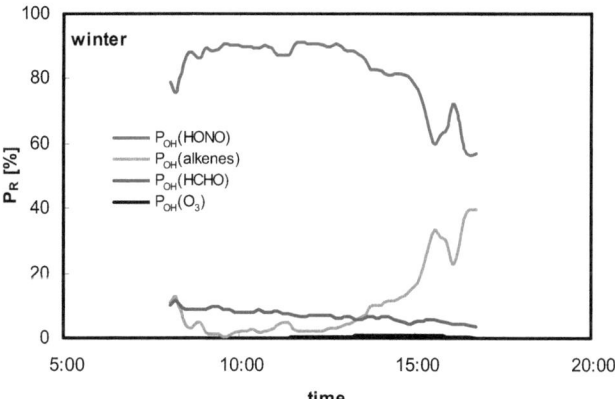

Fig. 3.2.13: Average relative diurnal daytime radical production rates (P_R) during winter in Santiago.

On a 24 h basis, $P_{OH}(HONO)$ was also the dominant radical source during summer and winter contributing ~52 % and ~81 % to P_R followed by P_{OH}(alkenes) (~29 %, ~12.5 %), $P_{OH}(HCHO)$ (15 %,~6.1 %) and $P_{OH}(O_3)$ (4 %, ~0.4 %), respectively.

The photostationary state concentration of HONO, $[HONO]_{PSS}$, defined as:

$$\text{E 3.9} \quad [HONO]_{PSS} = \frac{k_{OH+NO}[OH][NO]}{j(HONO) + k_{OH+HONO}[OH]},$$

was found to account for 69 % of the measured HONO on average during daytime in summer in comparison to 62 % during winter. Thus, the high daytime winter average absolute contribution of HONO in comparison to that of the summer campaign is due to a higher HONO concentration and the lower $[HONO]_{PSS}$ level in winter compared to that during summer.

During almost the entire daytime the HONO photolysis contribution was higher than any other primary source except in the early evening when the contribution from alkene ozonolysis starts to dominate. This is caused by the decreasing light intensity while the ozone and alkenes concentrations remaining high. In the early morning, the photolysis of HONO is the dominant source of the total radical budget. This is due to its low dissociation energy threshold and the high concentrations accumulated during nighttime.

The summer high morning peak production rate that slows down during the day has been previously observed in Los Angeles, Milan, Pabstthum (downwind of Berlin) and Mexico City (George et al., 1999; Alicke et al., 2002, 2003; Volkamer et al., 2007, respectively). However, in contrast to these studies, where the net OH production was very low in the afternoon, the relative contribution of the OH production by HONO photolysis never falls below 40 % for Santiago (see Fig. 3.2.11). This high daytime contribution of HONO is in good agreement with other recent studies under urban conditions (Ren et al., 2003; Acker et al., 2006b). The reason for the difference between the two sets of studies in which the contribution of HONO to afternoon radical production is either significant or negligible is still unclear. One potential explanation would be an overestimation of HONO due to interferences and sampling artefacts for all studies, in which wet chemical instruments were used (see Kleffmann and Wiesen, 2008). However, the LOPAP instrument used in the present study corrects for interferences and was successfully validated against the DOAS technique in a recent urban study in Milan (Kleffmann et al., 2006). In addition, a simple PSS analysis of the HONO data from the Milan campaign showed that HONO was also a strong net source of OH radicals during daytime, a result confirmed by the parallel co-located DOAS measurements (Kleffmann et al., 2006). This result is in contradiction to other DOAS measurements carried out at the same place under similar meteorological conditions and time of the year (Alicke et al., 2002). The reason for this difference is still unclear.

Another explanation for the different daytime contributions of HONO to OH radical formation in different studies may be the different sampling altitudes and strong vertical gradients during daytime. However, in the study of Alicke et al. (2002) the light path of the DOAS was even lower than the sampling height during the present study and no gradients were observed during daytime (Stutz et al., 2002). This is confirmed by other recent daytime gradient measurements, for which only small gradients were observed in the boundary layer during daytime (Kleffmann, 2007, Zhang et al., 2009). In addition, in the present study no horizontal gradients were observed towards the wall of the building on which the external sampling unit was fixed, excluding strong local wall sources. In conclusion, the reason for the different daytime contribution of HONO to the OH production in different studies remains unclear. The high contribution of HONO observed in the present study may be explained by the unique geographical situation of Santiago leading to very high pollution levels.

The relative diurnal contribution of HONO photolysis during winter is characterized by a high continuous contribution from the early morning till afternoon in contrast to that of the summer and the other studies mentioned above, in which the HONO contribution decreased to reach a minimum during the afternoon.

The lower contribution from the other sources during winter can be explained by the seasonal dependence of the radical precursors (see sec. 3.1.2). The low average daytime concentration of O_3 and HCHO is mainly due to the lower oxidation capacity during the winter. O_3 is basically a photochemical product while a substantial fraction of the measured HCHO in summer during afternoon has been shown to be photochemically produced (see sec. 3.2.9.1). Thus, the lower absolute and relative contribution of O_3 and HCHO to the radical initiation, P_R is mainly due to their lower photochemical formation rates and photolysis frequencies, $j(O^1D)$ and $j(HCHO)$, respectively during winter in comparison to those during summer (see sec. 3.2.9.1). In addition, the absolute water concentration in winter was ~20 %, lower than in summer which may also lower the $P_{OH}(O_3)$ during winter. The OH production from O_3 is ~27 times lower in winter than in summer which is quite comparable to that of ~25 times lower reported for New York City (Ren et al., 2006). The lower ozone level during winter also leads to much lower radical production from alkene ozonolysis. As a result of the higher HONO contribution during winter and the lower contribution from the other sources, net radical initiation, P_R was 10 % higher in winter than in summer.

Accordingly, the seasonal variation of the concentration of the radical precursors and their photolysis frequencies does have an impact on their relative contribution to the radical budgets. The order of relative importance of the radical precursors in winter in Santiago is similar to that in New York of HONO, namely 48 %, alkene ozonolysis, 36 %, photolysis of HCHO, 6 % and O_3, 1 % (Ren et al., 2006). However, due to the similar measured HONO levels and higher photolysis frequency in summer, HONO photolysis was more important in summer (56 %) than in winter in New York City (Ren et al., 2006).

3.2.7   Simulated OH Levels

The average diurnal profile of the OH concentration calculated by both, the MCM and PSS models during summer and winter are shown in Fig. 3.2.14 and Fig. 3.2.15, respectively. The maximum estimated OH concentrations calculated by MCM of 2.9×10^6 molecules cm^{-3} in winter is ~5 times lower than that of 1.4×10^7 molecules cm^{-3} during summer. Similarly, Ren et al. (2006) reported measured average maximum OH and HO_2 levels during winter of 1.4×10^6 molecules cm^{-3} and 0.7 pptv, respectively, one fifth the levels seen during the summer in New York City. However, Emmerson et al. (2005b) reported only a factor of 2 times lower OH levels during winter than in summer with HO_2 levels similar in summer and winter in Birmingham city centre. The maximum OH in winter occurs approximately 30 min after the maximum in $j(O^1D)$ which is shorter than that of ~1 h during summer. Similar seasonal variation of the gap between the maxima in [OH] and $j(O^1D)$ can be also observed in the study of Ren et al. (2006) in New York. Using different simplified photo-stationary state approaches, Rappenglück et al. (2000) and Rubio et al. (2005) estimated much lower OH concentrations of $~2.6 \times 10^6$ and $~8.8 \times 10^6$ molecules cm^{-3}, respectively. Possible reasons for these differences are that Rappenglück et al. (2000) did not consider HONO photolysis and alkenes ozonolysis, while Rubio et al. (2005) did not consider alkenes ozonolysis.

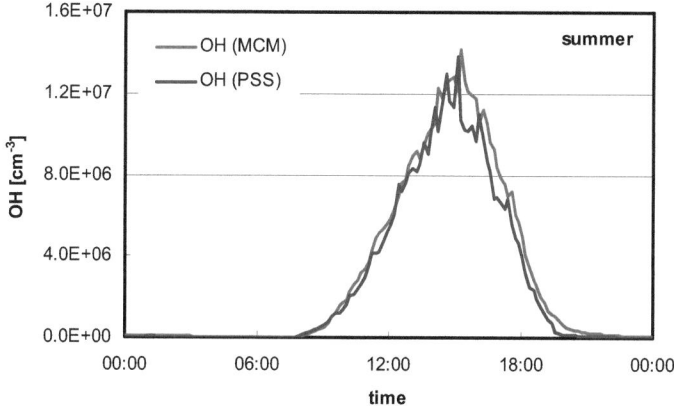

Fig. 3.2.14: Average diurnal variation of the OH concentration calculated by the MCM and PSS during summer.

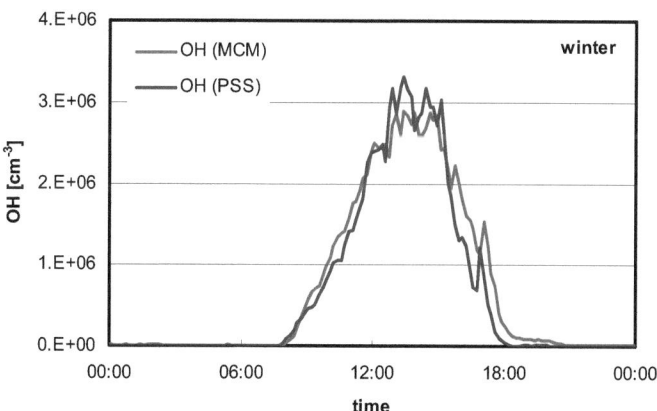

Fig. 3.2.15: Average diurnal variation of the OH concentration calculated by the MCM and PSS during winter.

The excellent agreement between the OH concentration profiles evaluated by both, the MCM and PSS models during both, summer and winter (Fig. 3.2.14 and Fig. 3.2.15, respectively) shows that the major OH radical sources and sinks are incorporated in the PSS model and that the secondary sinks (L_{OH}(OH+VOC)) are balanced with the secondary sources (P_{OH}(HO$_2$→OH)), which is in excellent agreement with the conclusion given at the end of sec. 3.2.5.

3.2.8 Correlation of OH and P$_R$ with j(O^{1}D) and j(NO$_2$)

In spite of the complexity of the mechanisms controlling OH concentrations, the OH correlation with j(O^{1}D) has shown to have a linear pattern in both, urban and rural environments and for long and short time periods (Rohrer et al., 2006, Kanaya et al., 2007). For Santiago, the calculated OH radical concentration in summer also correlates with j(O^{1}D) (R^2= 0.54), and j(NO$_2$) (R^2 = 0.56). An even stronger correlation between the measured daytime OH and j(NO$_2$) has been obtained in other studies (e.g. Kanaya et al., 2007 and references therein).

In addition, a better correlation between the total rate of radical initiation, P$_R$, and j(NO$_2$) compared to the correlation with j(O^{1}D) was observed in the present study during summer, especially for low j-values in the morning and evening (see Fig. 3.2.16 and Fig. 3.2.17). For high SZA in the morning and evening long wavelength radiation (j(NO$_2$)) increases much faster compared to short wavelength radiation (j(O^{1}D)). The good correlation of P$_R$ with j(NO$_2$) for high SZA demonstrates the importance of the UV-A rather than UV-B region for the production of OH during daytime, which is dominated by the daytime production of HONO.

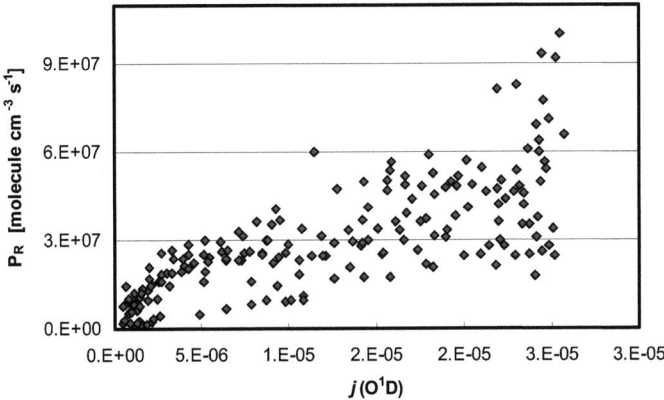

Fig. 3.2.16: Correlations between the total OH initiation rate, P$_R$ and j(O^{1}D).

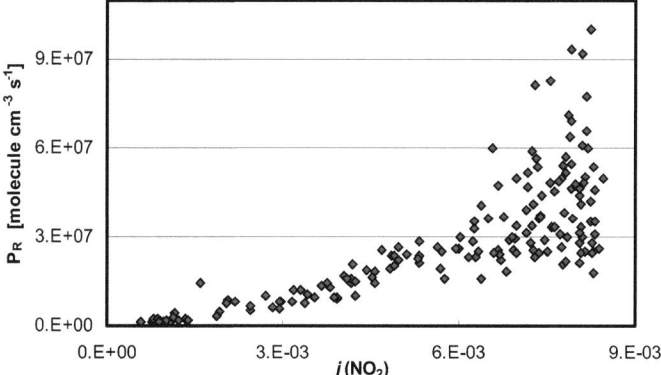

Fig. 3.2.17: Correlations between the total OH initiation rate, P_R and $j(NO_2)$.

3.2.9   Source Apportionments of the main OH Radical Precursors

As aforementioned, the OH radical initiation is controlled by the net radical sources, namely, photolysis of HONO, HCHO, O_3 and the reaction of alkenes with O_3. Source apportionment analyses of these primary sources in summer are discussed in this section except O_3, which is discussed in more details in sec. 3.3.

3.2.9.1   Formaldehyde (HCHO) Contribution

HCHO is a main photochemical oxidation precursor contributing ~16 % of the total primary radical sources, P_R, during daytime in Santiago. HCHO is both, primarily emitted and produced photochemically from the oxidation of VOCs (Friedfeld et al., 2002; Garcia et al., 2006). In this study, O_3 and NO_x were used as HCHO tracers for which NO_x has been assumed to be an indicator for primary HCHO resulting from direct emissions and O_3 as a photochemical indicator. The measured HCHO was described by the following relationship:

E 3.10 $[HCHO]_{measured} = \beta_o + \beta_1 \times [O_3] + \beta_2 \times [NO_x]$,

where β_o is the background HCHO (BKG), which stands here for the residual fraction of HCHO that cannot be accounted as photochemical or primary, and the factors β_1 and β_2 are the average weighted slopes of HCHO to O_3 and NO_x, respectively. For the whole summer campaign, values of $\beta_1 = 0.062$ and $\beta_2 = 0.018$ ppbv/ppbv, respectively were determined.

The photochemically formed HCHO (PHOT) comprises up to 70 % of the observed HCHO in the afternoon (Fig. 3.2.19). In contrast, during the early morning rush hour the primary HCHO (traffic) comprised up to 90 % (Fig. 3.2.19). Averaged on a daily basis, ~34 % of the measured HCHO is due to direct emissions while photochemical and background HCHO account for ~28 and ~38 %, respectively. The value of the direct emitted fraction is very similar to the 32±16 % previously reported by Rubio et al. (2006) while the sum of the photochemical and background fractions is similar to the secondary fraction reported during the summer in Santiago of 79±23 % (Rubio et al., 2006) and London, 74 % (Harrison et al., 2006). Since only 28 % of the HCHO is photochemically formed as a result of hydrocarbon oxidation, HCHO was considered as a net source of HO_2 (P_{HO_2}(prim)) in the present study.

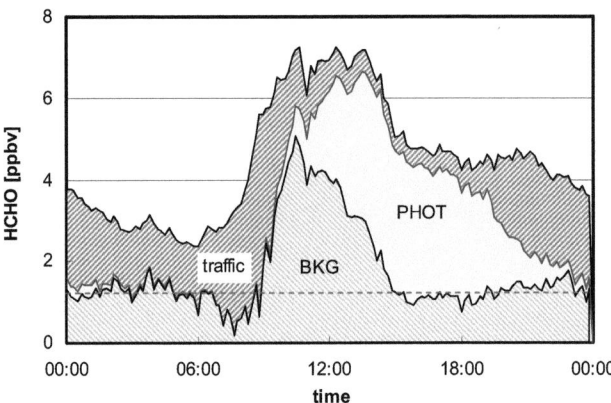

Fig. 3.2.18: HCHO source apportionment in Santiago, labelled as photochemically formed HCHO (PHOT), primary emitted HCHO (traffic) and the unaccounted sources (BKG). The red dotted line accounts for the baseline of the HCHO background during the early morning and afternoon (see sec. 3.2.9.1).

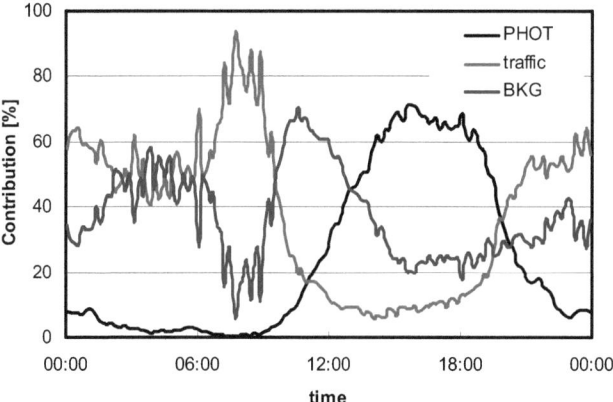

Fig. 3.2.19: Relative contributions of the HCHO sources in Santiago, labelled as photochemically formed HCHO (PHOT), primary emitted HCHO (traffic) and the unaccounted sources (BKG).

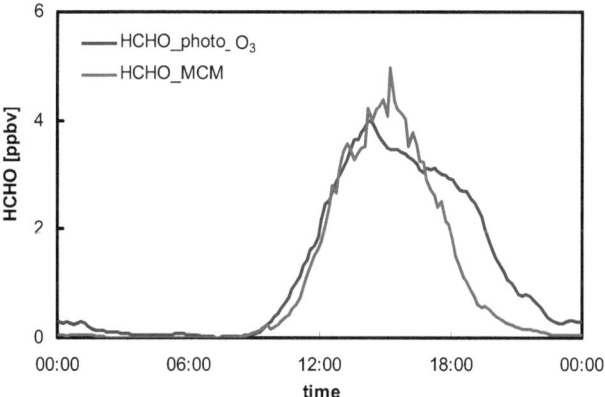

Fig. 3.2.20: Photochemical HCHO simulated with MCM and calculated with O_3 tracer.

Photochemical HCHO production has also been simulated using the MCMv3.1 photochemical box model constrained with all measured trace gases including NMHCs except measured HCHO. The photochemical HCHO calculated using O_3 as tracer matched well that calculated by the MCM model with a gap in the late afternoon (see Fig. 3.2.20). This gap however, is due to the afternoon ozone shoulder, which has become a typical feature during photochemical smog episodes in Santiago de Chile (Rappenglück, 2000, 2005).

Primary HCHO starts to build up in the early morning at about 6:30 h, nearly one hour before sunrise, and becomes the dominant source until ~9 h. The photochemical formation of HCHO follows the light intensity, and starts to increase nearly an hour after sunrise, becoming dominant at around ~13 h and reaching a maximum at ~16 h nearly 3 hours after the maximum in $j(NO_2)$. The photochemical HCHO contribution starts to decline at ~19 h, about 3 hours after the $j(NO_2)$ starts decreasing, while the primary HCHO turns again to be the dominant source until 2 h due to nighttime emissions.

The average background baseline of HCHO is less than 2 ppbv representing about 20 % of the total HCHO throughout the day (red dotted line in Fig. 3.2.18). The baseline of the background determines the average HCHO background values during the early morning and afternoon and is in agreement with that of Mexico City (Garcia et al., 2006). However, unexpected high background concentrations of HCHO, reaching a maximum of up to 5 ppbv at ~10 h, have been evaluated. One explanation is an underestimation of photochemical produced HCHO by the use of O_3 as tracer, since photochemically produced O_3 is efficiently titrated by the morning rush hour NO. In this case, photochemical HCHO would become more important after ~9 h. The use of O_x (O_3+NO_2) as tracer was not possible, since NO_2 is also linked to direct emissions (Carslaw and Beevers, 2005). Another explanation for the high HCHO background peak may be direct HCHO emissions that are not traced by NO_x (Garcia et al., 2006). These emissions should then however, be limited to the time period 9 - 14 h (see Fig. 3.2.18), which is unreasonable. Finally, the high background HCHO could also be caused by mixing of surface air masses with the residual layer in the morning when the boundary layer height is increasing. The concentration of HCHO in the residual layer could remain high from the previous day. Rappenglück et al. (2005) has also observed a similar background carbonyl peak at noontime in Santiago.

The contribution of each of the VOC classes (alkenes, alkanes, aromatics) to the photochemically formed formaldehyde has been calculated by the MCM model. As expected, the alkenes are the dominant photochemical HCHO precursor contributing alone more than 70 %, followed by aromatics, 18 %, and alkanes, 12 %. These contributions are in good agreement with those reported in Mexico City (Volkamer et al., 2007). From the alkenes, oxidation of isoprene contributes alone about 23 % to the photochemically produced HCHO, propene 11 % and α-pinene 9 %. From the aromatics, 1,3,5-trimethylbenzene represents 6 % followed by o-xylene, 4 %, and toluene, 3 %. From the alkanes, 2-metylbutane, decane and 3-methylpentane are the major sources contributing to about 3 %, 2 % and 1.6 % respectively.

OH is the dominant oxidant responsible for nearly 85 % of the total HCHO produced by the oxidation of hydrocarbons followed by alkene ozonolysis, 14 %. The contribution of NO_3 was found to be negligible.

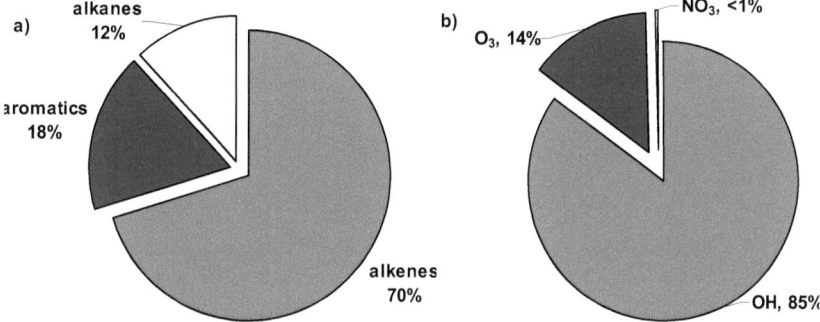

Fig. 3.2.21: Contribution of different a) hydrocarbons categories and b) oxidants to the photochemical formation of HCHO in Santiago during summer.

Table 3.2.1: The contribution of the different VOCs to the photochemically formed HCHO during summer in Santiago de Chile.

VOC	% contribution to HCHO
propene	11.0
1,3 butadiene	2.00
t-butene	2.20
cis-butene	1.40
3-methyl-1-butene	0.90
1-pentene	1.80
isoprene	23.0
t-pentene	2.60
cis-pentene	2.18
2-methyl-2-butene	6.92
1-hexene	2.08
2,3-dimethyl-1-butene	5.56
α-pinene	8.74
benzene	0.07
toluene	3.13
ethylbenzene	0.73
o-xylene	4.02
n-propylbenzene	0.17
1,3,5-trimethylbenzene	6.10
4-ethyltoluene	0.46
1,2,3-trimethylbenzene	1.77
styrene	1.37
propane	1.10
i-butane	0.68
n-butane	1.29
i-pentane	3.01
3-methylpentane	1.57
2-methylhexane	0.54
n-heptane	0.84
n-octane	0.59
decane	2.20

3.2.9.2 Alkene Ozonolysis Contribution

Unlike other OH radical sources, alkene ozonolysis can occur at night as well as during the day (Paulson and Orlando, 1996; Johnson and Marston, 2008). In this study, the ozonolysis of alkenes was found to be the second most important radical initiation source after HONO photolysis, accounting for 29 % of the OH formed during summer on a 24-h basis.

Although their total concentrations are only ~19 % of the total measured alkenes, internal alkenes contribute 86 % to the total alkene OH radical production given by $\Sigma k_{O_3+\text{alkene}}[\text{alkene}][O_3]\Phi_{OH}$ (see sec. 2.3.1), and nearly 21 % to the total primary radical production, P_R, as shown in Fig. 3.2.22. The order of efficiency in OH production from the reactions of ozone with alkenes is:

internal alkenes > cycloalkenes > terminal alkenes.

Among the internal alkenes, 2-methyl-2-butene and 2,3-dimethyl-2-butene have the highest contributions to the alkenes OH radical production with 37 % and 33 %, respectively (see Fig. 3.2.23). Cycloalkenes are represented by α-pinene alone and contribute about 6.6 % to the total alkene concentration, ~9 % to total alkene OH production and ~2 % to P_R. The other measured cycloalkenes are not yet included in the MCM. Terminal alkenes, while representing 75 % of the alkenes concentration, contribute only ~5 % to the total alkene OH production rate and about 1 % to P_R (Fig. 3.2.22).

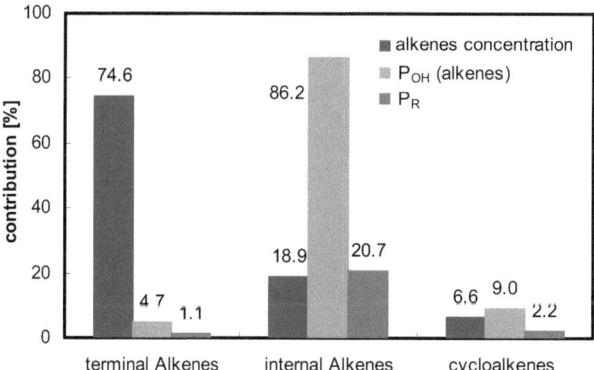

Fig. 3.2.22: Contributions of different alkenes categories to the net OH production.

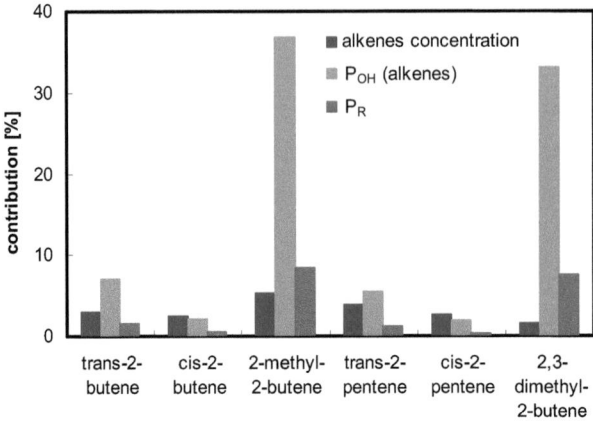

Fig. 3.2.23: Contributions of internal alkenes to the net OH production.

3.2.9.3 HONO Daytime Contribution

As already discussed, over the last few years it has been demonstrated that the contribution of nitrous acid to the primary radical production, P_R, has been frequently underestimated (e.g. Ren et al., 2003; Kleffmann et al., 2005; Acker et al., 2006a). High measured daytime concentrations point to an additional strong HONO source (Kleffmann, 2007), for which several photochemical reactions have recently been proposed from laboratory studies (Zhou et al., 2003; George et al., 2005; Bejan et al., 2006; Stemmler et al., 2006, 2007, Gustafsson et al., 2006; Ndour et al., 2008).

On average, [HONO]$_{PSS}$ (see sec. 3.2.6) was found to account for about 69 % of the observed HONO concentration in summer reaching its maximum contribution during the rush hour peak time at ~10 h coinciding with the NO peak. During the early afternoon (12:30 – 15:00 h), when the absolute production rate of OH by HONO photolysis was highest, the PSS contributed on average ~66 % of the measured HONO. Thus, one reason for the extreme high HONO daytime concentrations observed is the daytime production of HONO by the gas phase reaction NO+OH caused by the very high levels of OH and NO. However, this reaction and the uncertainty in the PSS concentration by only gas phase chemistry (see below) cannot explain the measured daytime values of HONO alone. If the heterogeneous dark conversion of NO_2 (see sec. 3.2.9.4) is included, the PSS increases by only 4 % during noon; thus, an additional daytime source of HONO is needed. The most important uncertainty in the calculation of the PSS concentration besides the measured [NO], [HONO] and j(HONO) values is the modelled [OH]. However, an average maximum OH concentration of 2.2×10^7 molecules cm^{-3}, which is about 155 % of the modelled OH, is required to get [HONO]$_{PSS}$ equal to measured values. The OH simulated by the MCM model was validated through different field intercomparisons and showed excellent agreement with that measured, especially under such high NO$_x$ conditions (Mihelcic et al., 2003; Sheehy et al., 2008). Recently, Sheehy et al. (2008) reported maximum OH over prediction by the MCM of 20 % during afternoon. In contrast, for Santiago an under prediction of the modelled OH level by ~55 % would be necessary to explain the daytime concentrations of HONO. Therefore, additional average daytime HONO sources of 1.7 ppbv h^{-1} are necessary to explain observed HONO levels during summer.

These additional daytime HONO sources become obvious from the diurnal variation of the HONO/NO$_x$ ratio (Fig. 3.1.4, p. 58). While the night-time behaviour, with a linear increase of the HONO/NO$_x$ ratio from 2-5 %, is typical for urban conditions and can be explained by known emission and heterogeneous conversion of NO$_2$ on ground surfaces (Alicke et al., 2002; Kleffmann et al., 2002, 2003; Vogel et al., 2003), the second daytime maximum, reaching almost 8 %, has not been observed in our previous urban studies in such a pronounced manner. A daytime maximum under urban conditions was however observed for the city of Rome (Acker et al, 2006b) and is also typical for remote and mountain site measurements (see e.g., Huang et al., 2002; Kleffmann et al., 2002; Acker et al., 2006a, Kleffmann and Wiesen, 2008). The daytime maximum in the HONO/NO$_x$ ratio can only be explained by a very strong additional photochemical HONO source.

Three photochemical mechanisms were identified recently, two of them being well correlated to j(NO$_2$) (George et al., 2005; Bejan et al., 2006; Stemmler et al., 2006, 2007; Gustafsson et al., 2006; Ndour et al., 2008, Li et al., 2008), while the photolysis of nitric acid (Zhou et al., 2003) would better correlate to j(O^{1}D), caused by the much lower wavelength range of the nitric acid photolysis. This was tested by plotting the campaign averaged net production rate of OH radicals due to HONO photolysis against j(NO$_2$) and j(O^{1}D). Both plots (j(NO$_2$), $R^2 = 0.62$ and j(O^{1}D), $R^2 = 0.45$) show that the daytime source is correlated with the light intensity, confirming former assumptions of a photochemical production of HONO. However, since a better correlation was obtained when $j(NO_2)$ was used, especially for low j-values, the heterogeneous conversion of NO_2 on photosensitized organics (George et al., 2005; Stemmler et al., 2006, 2007) and gas phase photolysis of organic nitrogen compounds (e.g. nitrophenols, Bejan et al., 2006) in addition to the reaction of electronically excited NO_2 with water vapour (Li et al., 2008, see sec. 1.1.1) may be of higher importance compared to the nitric acid photolysis in Santiago. Similar results were obtained when plotting P_R against $j(NO_2)$ and $j(O^1D)$ (see Fig. 3.2.16 and Fig. 3.2.17, sec. 3.2.8).

3.2.9.4 HONO Dark Sources:

Besides photochemical daytime sources of HONO, formation of HONO during the night by heterogeneous conversion of NO_2 on humid surfaces is well known (Alicke et al., 2002). The nighttime accumulated HONO is of major importance to initiate radical chemistry in the morning after sunrise. The dark heterogeneous rate constant of HONO formation, k_{het}, due to the first order conversion of NO_2 on humid surfaces ($NO_2 + X \rightarrow HONO$) has been estimated from the increase of the $HONO/NO_2$ ratio during the night (see also Alicke et al., 2002). An average k_{het} of $(3.5\pm1.9)\times10^{-6}$ s^{-1} has been obtained, which is similar to that of $(3.3\pm1.4)\times10^{-6}$ s^{-1} obtained by Alicke et al. (2002). This heterogeneous rate constant has been found to correlate inversely with the wind speed ($R^2 = 0.65$) confirming heterogeneous formation on ground surfaces during the night (Kleffmann et al., 2003). However, almost no correlation of k_{het} with relative humidity was observed ($R^2 = 0.086$) in contrast to the study by Stutz et al. (2004). The lack of water dependence can be explained by the heterogeneous conversion of NO_2 into HONO on adsorbed organics (Arens et al., 2002; Gutzwiller et al. 2002; Ammann et al., 2005), which are persistent on any urban surface. For this type of reaction only moderate humidity dependence was observed in the laboratory (Arens et al., 2002) for a humidity range comparable to the present study. In addition, NO_2 conversion on organic surfaces (Arens et al., 2002) is much faster than the typical proposed reaction of NO_2 with water on surfaces (Finlayson-Pitts et al., 2003) at atmospheric NO_2 levels and thus is a more reasonable source for nighttime formation of HONO in the atmosphere.

3.2.10 Conclusion

The oxidising capacity of the atmosphere has been studied for the first time over the urban area of Santiago, Chile, during two extensive measurement campaigns in the summer and winter 2005. The effects of seasonal changes on the radical budgets have been subjected for a detailed analysis. A zero dimensional photochemical box model based on the MCMv3.1 and constrained to a suite of ancillary measurements was used to calculate the concentration and radical budget of the radical species, OH, HO_2 and RO_2. In addition, a PSS assumption was used to analyse the main radical sources and sinks.

Total production/destruction rates of the OH and HO_2 in winter are ~2 times lower than that during summer which is mainly due to the high NO_x levels and the lower photochemical activity during winter. Production and destruction of HO_2 was dominated by the peroxy radical recycling reactions, RO_2 and HO_2 with NO with the later comprising 80 % and 66 % of the OH production rate during summer and winter, respectively. OH loss was dominated by its reaction with hydrocarbons which contribute 79 % and 67 % of the total OH loss during summer and winter, respectively.

The average daytime OH reactivity during winter is ~2 times higher than in summer which is due to apparent elevated pollutant levels during winter in comparison to that during summer. The daytime average lifetime of OH, τ_{OH} due to reaction with NO_2 is 0.07 s in comparison to that of 0.24 s during summer. The lower total production and destruction rates of OH during winter are a result of the shorter lifetime of OH during winter owing to the higher NO_x levels during winter. The low average HO_2/OH ratio of 9 and 8 in winter and summer, respectively, is typical for highly polluted environments and implies a high recycling efficiency towards OH while the RO_2/HO_2 ratio of 1-1.5 for both summer and winter is comparable to that of other urban studies.

During both summer and winter, there was a balance obtained between the secondary production, $P_{OH}(HO_2 \rightarrow OH)$ and destruction $L_{OH}(OH + VOC)$ rates of OH which has been further investigated for other studies using a balance ratio, BR. Interestingly; a BR value of 1.0 was obtained from the slope of the correlation between the secondary production and destruction of OH for all field studies used for comparison. Therefore, initiation sources of HO_2 and RO_2 should not be considered as net OH sources under urban conditions.

The excellent agreement between the OH concentration profiles evaluated by both, the MCM and PSS models during summer and winter shows that the major OH radical sources and sinks are incorporated in the PSS model and that the secondary sinks, L_{OH}(OH + VOC) are balanced with the secondary sources P_{OH}(HO$_2$→OH). For daytime conditions during summer and winter, HONO photolysis has the highest contribution (55 %, 84 %) followed by alkenes ozonolysis (24 %, 9 %), HCHO photolysis (16 %, 6.5 %) and ozone photolysis (5 %, 0.5 %) to OH radical initiation rate. High net mean and maximum OH production rates by HONO photolysis of 2.9 ppbv h^{-1} and 6.2 ppbv h^{-1} have been determined during winter, which is much higher than those of 1.7 ppbv h^{-1} and 3.1 ppbv h^{-1}, respectively, determined during summer.

The modelled high OH concentrations during summer show that the high daytime concentrations of HONO cannot be explained by known gas phase chemistry and suggest the presence of an additional strong daytime source of HONO. This conclusion is further supported by the observation of a second daytime maximum in the HONO/NO$_x$ ratio that is higher in summer (higher photochemical activity) than in winter. The better correlation of the daytime HONO source with j(NO$_2$) compared to j(O^{1}D) shows that the recently proposed nitric acid photolysis channel cannot be explain daytime HONO formation. The major contribution of HONO to the direct OH radical production is in good agreement with several recent studies and highlights the importance of HONO measurements in studies which focus on the radical chemistry of the atmosphere.

Alkene ozonolysis represented the second most important direct source of OH radicals with internal alkenes having the major contribution to OH radical formation. HCHO source apportionment has been achieved using NO$_x$ and O$_3$ as direct emission and photochemical tracers, respectively. Photochemical HCHO comprises up to >70 % of the observed HCHO during the afternoon. The HCHO photochemical source apportionment has revealed that alkenes contribute most by 70 % followed by aromatics, 18 %, and alkanes, 12 %.

Generally, the balance between secondary radical formation and destruction was mainly due to the high recycling efficiency owed to the high NO$_x$ conditions experienced in Santiago. This high NO$_x$ conditions has been further evidenced in the next section where a detailed analysis of the O$_3$-VOC-NO$_x$ sensitivity was performed.

3.3 Summertime Photochemical Ozone Formation in Santiago, Chile

In the present study, the O_3-VOC-NO_x sensitivity has been analyzed by simulating photochemical ozone formation under different VOC/NO_x regimes using a photochemical box model based on the Master Chemical Mechanism, MCMv3.1 (see sec. 2.3.2.2). In addition, a set of potential empirical indicator relationships has been used to determine whether ozone formation is VOC or NO_x limited. The MCM box model has also been used to determine the potential of each measured hydrocarbon to form ozone under representative summertime conditions often experienced in Santiago. The measured ozone is henceforth referred to as "ozone" unless additionally labelled by "modelled".

3.3.1 Simulated Ozone Levels and Model Limitations

The simulated and measured ozone concentrations for the average campaign data are shown in Fig. 3.3.1, whereas Fig. 3.3.2 shows only those for March 18, 2005. For both, the average day campaign and the March 18, the simulated maximum ozone concentration (base scenario) was higher than the observed ozone by ~40 %, as shown in Fig. 3.3.1 and Fig. 3.3.2. In contrast, in the morning and evening, measured ozone exceeded modelled values. Especially in the afternoon, measured ozone showed an afternoon shoulder at ca. 18:00 h, which has been already observed in other studies in Santiago (Rappenglück et al., 2000, 2005).

Different reasons may explain the deviation of measured and modelled ozone concentrations. Bloss et al. (2005a) evaluated the detailed aromatic mechanisms for MCMv 3.1 and found that peak ozone concentrations were significantly over-estimated for the substituted aromatics, which may also contribute to the over-estimation of maximum ozone levels seen here. In addition, since ozone formation is a non-linear photochemical process, the use of average campaign data rather than individual day measurements may also have an impact on the estimated ozone concentrations. However, the maximum simulated ozone concentration for March 18 (a day for which complete data were available) was also higher by ~40 % than the measured levels (Fig. 3.3.2). Accordingly, since the average day campaign has the advantage of representing the average summer time conditions, it has been further considered for sensitivity analysis (see below).

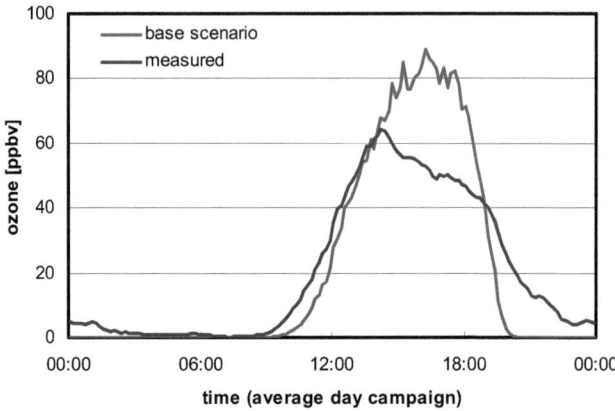

Fig. 3.3.1: Diurnal variation of measured and simulated ozone concentrations for the average day campaign.

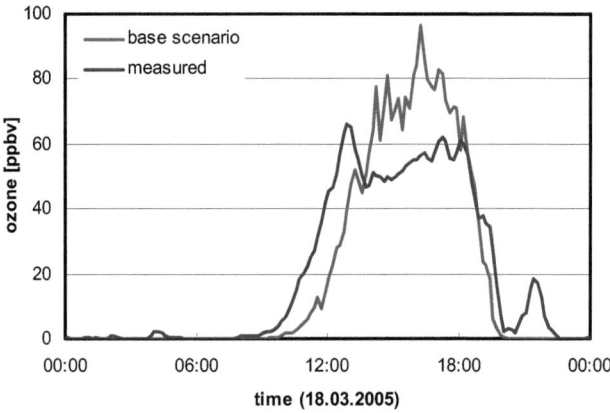

Fig. 3.3.2: Diurnal variation of measured and simulated ozone concentrations for March 18, 2005.

In addition, since photochemical ozone formation during the different sensitivity analyses was simulated based on a one-day model run, the simulated ozone concentration should be considered even as a lower limit. This is due to delayed formation of ozone (calculated as the difference between the maximum simulated ozone concentration in the first and fifth day simulation runs), which leads to an underestimation of maximum ozone levels using the one-day model run. Derwent et al. (2005) explained delayed ozone formation for alkenes and carbonyls and have shown that their degradation products (i.e. $\sum$PANs and aldehydes) formed on the first day may take part in the ozone formation over longer time scale in the model.

Furthermore, since not all the measured VOCs were constrained into the model, and since not all VOCs could be quantified by the used GC analysis technique employed, ozone formation will also be underestimated due to the underestimation of HO_2 and RO_2 that result from the oxidation of VOCs under VOC limited conditions.

The meteorological and topographical conditions are also important factors that govern air pollution in Santiago de Chile (Schmitz et al., 2005; Gramsch et al., 2006), however, are not considered in the used box model and thus may explain observed differences. The wind direction during the afternoon from 12:00 to 18:00 h is southwest with the wind speed reaching its maximum at ~18:00 h (Gramsch et al., 2006) carrying biogenic emissions to this part of the city. Thus, the lower local measured ozone levels during the early afternoon may have resulted from this dilution while the afternoon ozone shoulder observed at later time may be explained by the delayed ozone formation due to these biogenic emissions (see sec. 3.3.8). However, this wind profile leads also to increasing levels of pollutants including ozone in the eastern part of the city downwind the measurement site during afternoon (Schmitz et al., 2005, Gramsch et al., 2006).

Another possible reason for the overestimated ozone concentrations could be the depth of the PBL in the model. The boundary layer height strongly depends on the location and the daytime. However, since BLH measurements were not available during the campaign, we used the average maximum BLH of 1400 m (above ground), which lies within the range of that, previously reported during summer (Schmitz, 2005; Gramsch et al., 2006 and references therein). Since surface deposition depends on the boundary layer height in the model, a different BLH may also influence the modelled O_3.

In addition, the applied model is a zero dimensional photochemical box model and hence assumes that the box is continuously well mixed. Therefore, simulated ozone concentrations during the early and mid-morning may have been underestimated because the amount of ozone that may be entrained from aloft in the morning could not be simulated.

In conclusion, several reasons may explain differences in the modelled and measured ozone concentrations. However, because the ozone sensitivity analysis and the determination of the ozone formation potentials are basically a relative analysis (i.e. relative to the base model scenario, see sec. 2.3.2.2), the one-day model ozone simulations and the use of a box model is expected to be adequate. Similar photochemical box models have been also previously used for the analysis of ozone photochemistry and as a tool for assessing air quality in different urban areas (e.g. Huang et al., 2001 and references therein).

3.3.2   VOC/NO$_x$ Ratio

According to a report from the National Research Council (NRC, 1991), a VOC/NO$_x$ ratio lower than 10 correspond to VOC-sensitive ozone production, while VOC/NO$_x$ ratios >20 correspond to NO$_x$-sensitive ozone production. Using all quantified VOCs (from identified and unidentified hydrocarbons, see sec. 3.1), average summertime VOC/NO$_x$ ratio (ppbC/ppbv) of ~8 was calculated during the morning rush hour (~9 h) in Santiago. However, this ratio increases dramatically to reach ~27 at noon with an average diurnal value of ~14. In addition, since not all VOC could be quantified by the GC analysis applied in the present study, an even higher ratio should be considered. Previously, a high VOC/NO$_x$ ratio has been reported for Santiago and a NO$_x$ limited regime was accordingly assigned (Jorquera et al., 1998b). The VOC/NO$_x$ rule however, does not account for the impact of the VOC reactivity and has shown to fail in more sophisticated photochemical models (Sillman, 1999 and references therein). The VOC/NO$_x$ ratio thus may not correctly represent the photochemical ozone sensitivity of Santiago, which will be further investigated in this study using the MCM photochemical box model.

3.3.3   Ozone Sensitivity Analysis using the MCM Box Model

To avoid the ambiguity of the definitions and the high uncertainties associated with the use of the VOC/NO$_x$ ratio, the ozone sensitivity has been determined by simulating the photochemical ozone formation using the MCMv3.1 (see sec. 2.3.2.2) under different VOCs and NO$_x$ conditions (scenario_2). The simulated maximum ozone concentrations obtained from the 1-day and 5-day model runs at different VOCs and NO_x concentrations are shown in Fig. 3.3.3 and Fig. 3.3.4, respectively. Since both, the 1-day and 5-day model simulations give the same results for the selected sensitivity points as shown in Fig. 3.3.3 and Fig. 3.3.4, the 1-day sensitivity analysis was further considered due to its shorter integration time.

From the NO sensitivity analysis (Fig. 3.3.3), decreasing NO concentrations lead to an initial increase in the maximum ozone concentration followed by a decrease at 5 % of the measured NO concentration. The VOC sensitivity (see Fig. 3.3.4) shows that ~400 times the measured VOC concentration is required to reach the maximum simulated ozone level owing to the high NO levels observed. Thus, even if all VOCs would have been quantified and constrained in the model, the ozone sensitivity would remain VOC-limited. The initial increase in O_3 as NO decrease is caused by the reduced importance of the reaction of ozone with NO and the decreasing loss of OH radicals by the reaction with NO_2 irreversibly forming HNO_3. Higher OH radical concentrations lead to higher hydrocarbon oxidation rates and thus to increasing ozone formation rates. The sharp decrease of ozone as NO decrease below 5 % of the measured NO values (Fig. 3.3.3) is due to the NO_x limited conditions. Under these conditions, NO_2 production from the NO reaction with RO_2 would be limited by the NO concentration leading to a reduction in the ozone formation rate with further decreasing NO. Both, VOC- and NO_x-sensitivities clearly show that ozone formation in Santiago is VOC-limited. The VOC limited regime is typical for highly polluted environments and has been reported previously in different urban studies (e.g. Zhang et al., 2007, Lei et al., 2007 and references therein).

The diurnal variation of the simulated ozone profiles of the NO_x and VOCs sensitivity analysis is shown in Fig. 3.3.5 and Fig. 3.3.6, respectively (for clarity, not all are shown) and gives an indication about the ozone response to changes in the NO_x and VOC emissions. Reductions of NO levels by a factor of <20 (incremental factor >0.05) do not result in a considerable change in the timing of the ozone peak, while a slight delay of ~2 h is observed when using an incremental factor of <0.05, (see Fig. 3.3.5). In contrast, the ozone peak of the simulated ozone profiles of the VOC-sensitivity increased and was shifted to an earlier time by ~5 h when VOC levels were increased by more than a factor of 80 (see Fig. 3.3.6). This is due to the NO_x limited regime that has been reached under these conditions, for which O_3 formation follows the diurnal variation of NO.

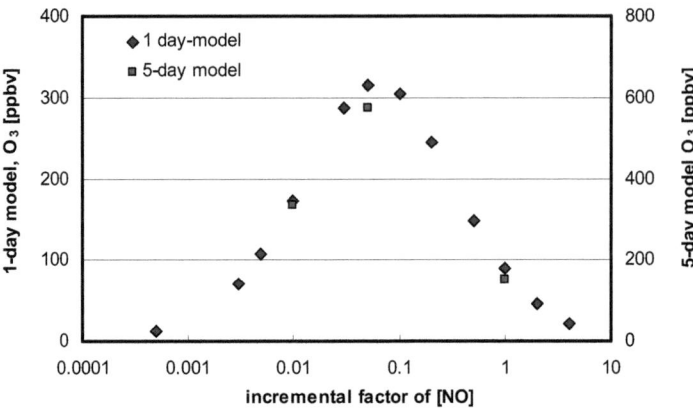

Fig. 3.3.3: MCM-NO_x sensitivity analysis shown as the maximum modelled O_3 concentration as a function of the concentrations of NO (incremental factor of 1 is equal to the measured concentration).

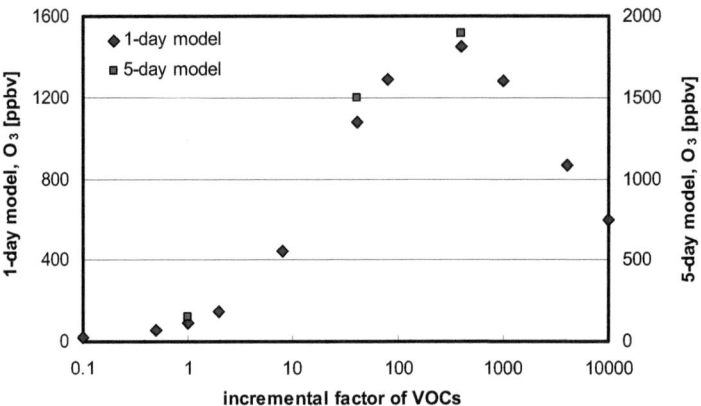

Fig. 3.3.4: MCM-VOC sensitivity analysis shown as the maximum modelled O_3 concentration as a function of the concentrations of VOC (incremental factor of 1 is equal to the measured concentration)

The enhancement of the ozone peak that is shifted to earlier time (Fig. 3.3.6) as a response to increasing the VOC levels and the initial increase in ozone levels as a response to reductions in NO levels (Fig. 3.3.5) both demonstrate the VOC-limited chemistry in Santiago.

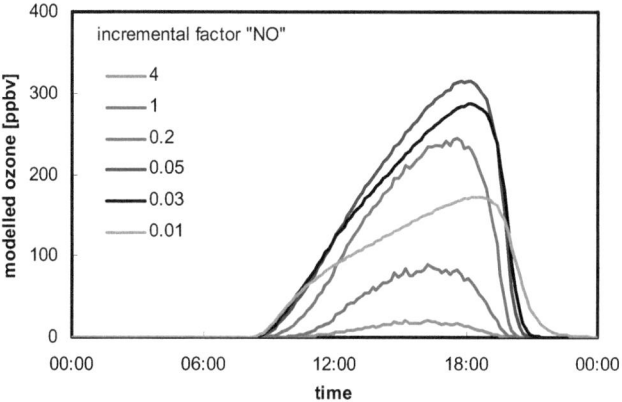

Fig. 3.3.5: Simulated ozone diurnal profiles for the NO-sensitivity study.

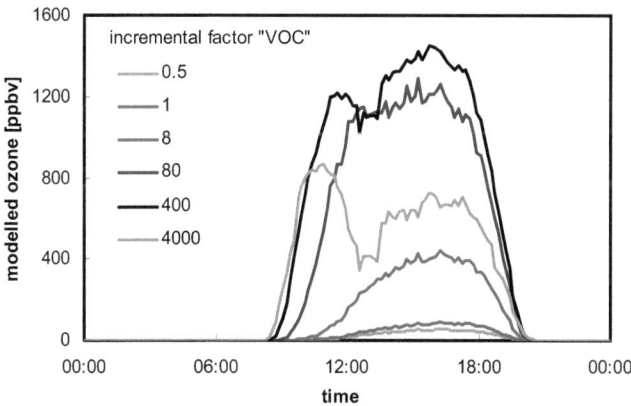

Fig. 3.3.6: Simulated ozone diurnal profiles for the VOC-sensitivity study.

3.3.4 Empirical Indicator Species/Relationships

The photochemical reactions leading to the formation of O_3 simultaneously produce a range of oxidised nitrogen species such as HNO_3, $HONO$, N_2O_5, $\Sigma RONO_2$ and $\Sigma PANs$. The sum of these simulated (HNO_3, $2N_2O_5$, $\Sigma RONO_2$ and $\Sigma PANs$ (except PAN)) and measured (HONO, PAN) photochemical species represent a large proportion of the NO_x oxidation products, NO_z ($NO_z = NO_y - NO_x$) and therefore was used to calculate NO_z. HNO_3 was found to account for ~52 % of NO_z during daytime (8-19 h) followed by $\Sigma RONO_2$, ~31 %, $\Sigma PANs$, ~9 % and HONO, ~8 % while N_2O_5 has a negligible contribution to NO_z. These oxidised nitrogen species, in addition to the modelled H_2O_2, can be used in photochemical indicator relationships in order to determine whether the atmospheric environment is NO_x- or VOC-sensitive (Sillman, 1995, Kleinman, 2000).

Using Eulerian model calculations, Sillman (1995) showed that the VOC-sensitive chemistry is linked to afternoon NO_y >20 ppbv, O_3/NO_z <7, $HCHO/NO_y$ <0.28 and H_2O_2/HNO_3 <0.4 (for the significance of these ratios refer to Sillman, 1995). NO_y was calculated as ($NO_y = NO_x + NO_z$). The average afternoon (14-18 h) values of these indicator relationships in Santiago are NO_y = 64 ppbv, O_3/NO_z = 1.41, $HCHO/NO_y$ = 0.08 and H_2O_2/HNO_3 = 0.014, respectively, confirming that ozone formation in Santiago is VOC limited, in agreement with the results from the O_3-VOC-NO_x sensitivity analysis (see sec. 3.3.3).

3.3.5 Ozone Production Efficiency, OPE

The ozone production efficiency, OPE is defined as the net number of molecules of O_3 generated per molecules of NO_x oxidized (e.g. Trainer et al., 1993). OPE has been obtained in this study from the correlation of O_x ($NO_2 + O_3$), representing the ultimate formation of ozone, against the sum of NO_x oxidation products, NO_z (cf. Finlayson-Pitts and Pitts, 2000, Rickard et al., 2002).

The linear correlation between O_x and NO_z during the ozone formation period from ~9-14 h (cf., Fig. 3.3.7) is given by the relation (O_x = (1.48±0.04) × NO_z + (23.0±1.1 ppbv) (R^2 = 0.98)). The intercept of the correlation between O_x and NO_z (23.0 ppbv) represents the ozone background level in Santiago. This value is towards the lower end of the current global ozone background range of 24-42 ppbv reported in other studies (e.g. see Finlayson-Pitts and Pitts, 2000 and references therein). The slope (~1.5) gives the OPE which is at the lower limit of the OPE range of 1.5-10 reported for different urban studies (Rickard et al., 2002 and references therein). The reason for the small slope is the high NO_x concentration in Santiago for which the reactions of (NO$_2$ + OH), (NO$_2$ + acyl peroxy radical (RC(O)OO)) and (NO + RO$_2$) to form HNO$_3$, $\sum$PANs and RONO$_2$, respectively, become important, which increases the NO$_z$ concentration and removes OH and NO$_2$ from the system, therefore decreasing the amount of O$_3$ formed.

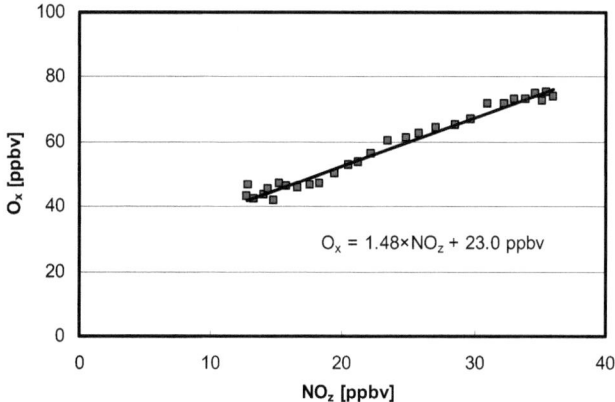

Fig. 3.3.7: Correlation between the average O$_x$ (NO$_2$ + O$_3$) and NO$_z$ concentrations during the ozone formation period (9-14 h).

3.3.6 Photochemical Ozone Production

The net instantaneous photochemical production rate of ozone or the ozone tendency, N_{O_3} is a measure of the ozone productivity of an air mass neglecting the deposition processes (Salisbury et al., 2002) and provides a pathway to understand high O$_3$ events (Kleinman et al., 2005). N_{O_3} has been calculated using the following equation (cf. Sheehy et al., 2008):

E 3.11 $N_{O_3} = P_{O_3} - L_{O_3}$,

where P_{O_3} and L_{O_3} represent the instantaneous ozone production and loss processes, respectively. P_{O_3} is given by:

E 3.12 $P_{O_3} = P_{O_3}(HO_2) + P_{O_3}(RO_2)$,

for which:

E 3.13 $P_{O_3}(HO_2) = k_{(HO_2+NO)}[NO][HO_2]$,

E 3.14 $P_{O_3}(RO_2) = \sum k_{(RO_2i+NO)}[NO][RO_2]_i$.

L_{O_3} is defined by:

E 3.15 $L_{O_3} = k_{(OH+NO_2)}[OH][NO_2][M] + P_{RONO_2}$.

During daytime (8 - 19 h), the average modelled P_{O_3} (RO_2) and P_{O_3} (HO_2) contribution to the instantaneous net ozone production rate, P_{O_3} are ~34 and ~28 ppbv h^{-1}, which correspond to 55 % and 45 %, respectively (see Fig. 3.3.8). The loss term, L_{O_3} in equation E 3.15 accounts on average for ~4.3 ppbv h^{-1} and thus is reducing N_{O_3} by ~7 %. This value is comparable to the ~10 % reported for Mexico City (Shirley et al., 2006) and strengthens the need to include this loss term when calculating the instantaneous ozone production rate. The HO_2 contribution to the instantaneous ozone production rate, P_{O_3} (HO_2) , reached a maximum of 71 ppbv h^{-1} at ~15 h (Fig. 3.3.8) which is higher than that of 48 ppbv h^{-1} calculated from the median measured HO_2 concentrations in Mexico City (Shirley et al., 2006).

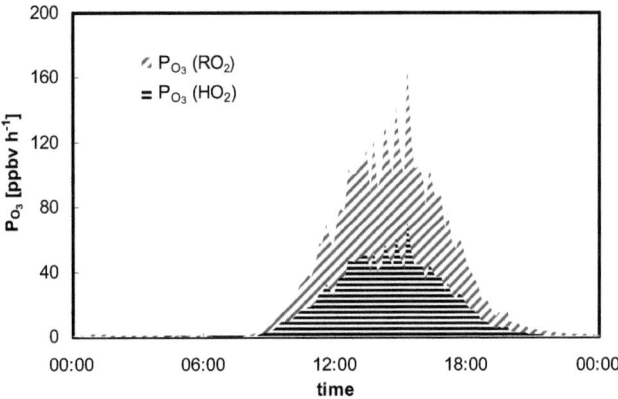

Fig. 3.3.8: Average diurnal profiles of HO_2 and RO_2 contribution to O_3 formation.

The average daytime N_{O_3} peaks at ~160 ppbv h^{-1} at ~15 h (see Fig. 3.3.9) and is higher than the modelled value of ~20 ppbv h^{-1} calculated for New York City (Ren et al., 2003) and 89 ppbv h^{-1} calculated using modelled RO_2 and measured HO_2 for Mexico City (Sheehy et al., 2008). Kleinman et al. (2005) reported N_{O_3} values that vary from 25 to 140 ppbv h^{-1} in different US urban areas. The average day total net instantaneous ozone production rate in

Santiago of ~700 ppbv day^{-1} is about 5 times higher than that modelled of 150±100 ppbv day^{-1} for New York City (Ren et al., 2003) and about 2 times higher than that of 319 ppbv day^{-1} for Mexico City (Shirley et al., 2006). The calculated ozone production rates in this study are based on modelled RO$_2$ and HO$_2$ concentrations. Therefore, they should be considered as a lower limit because not all the measured VOCs were constrained in the model and not all VOCs could be quantified. This will certainly lead to an under-estimation of the RO$_2$ and HO$_2$ concentrations as a result of the VOC-sensitive environment (see sec. 3.3.3) and consequently an under-estimation of the N$_{O_3}$. Similarly, Ren et al. (2003) reported higher ozone production from measured HO$_2$ than from that modelled at high values of NO.

It should be also mentioned that difficulties associated with the ambient measurements of HO$_2$ and RO$_2$ may also lead to significant uncertainties in the calculation of N$_{O_3}$. Due to difficulties in measuring RO$_2$, Sheehy et al. (2008) used modelled RO$_2$ and measured HO$_2$ to calculate N$_{O_3}$ (see above equations E 3.13+E 3.14) and considered the results also as lower limits because of the lower than expected RO$_2$/HO$_2$ ratio. Thus, they assumed much higher ozone production when this ratio approaches unity. Similarly, due to unexpected behaviour of the measured HO$_2$ diurnal profile, Ren et al. (2003) used the modelled HO$_2$ value to calculate the N$_{O_3}$ and suggested ozone production of a factor of 1.5 higher if measured HO$_2$ were used. Thus, using the modelled values of HO$_2$ and RO$_2$ is a justified alternative to calculate the lower limit of N$_{O_3}$ in this study.

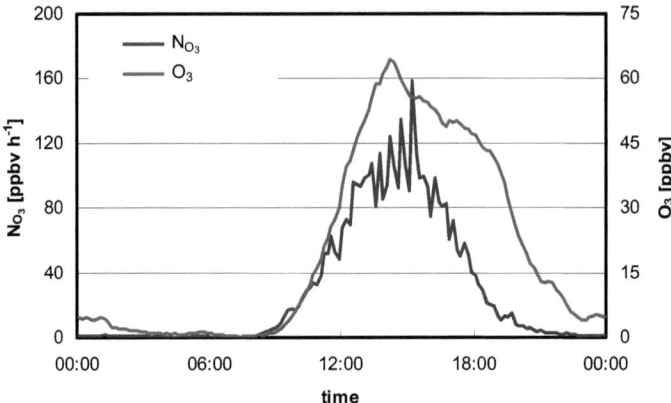

Fig. 3.3.9: Average diurnal profiles of the O_3 concentration and the net instantaneous production rate of O_3 (N_{O_3}).

The average diurnal profile of N_{O_3} that peaks at ~15 h (see Fig. 3.3.9) is similar to that observed in New York City (Ren et al., 2003) but different from that observed in Mexico City, which peaks during the mid-morning (Sheehy et al., 2008). The measured ozone profile tracks that of the ozone production rate reasonably well except for the late ozone afternoon shoulder (see Fig. 3.3.9). The low ozone production efficiency associated with high N_{O_3} is an apparent result of a highly polluted environment subjected to high NO_x levels (Rickard et al., 2002).

This result is consistent with the (anti-) correlation between N_{O_3} and NO shown in Fig. 3.3.10. Such conditions of high O_3 production associated with VOC-limited conditions are similar to those observed in Tokyo (Huang et al., 2001), Philadelphia and Phoenix, US (Kleinman et al., 2005) and Hong Kong (Zhang et al., 2007).

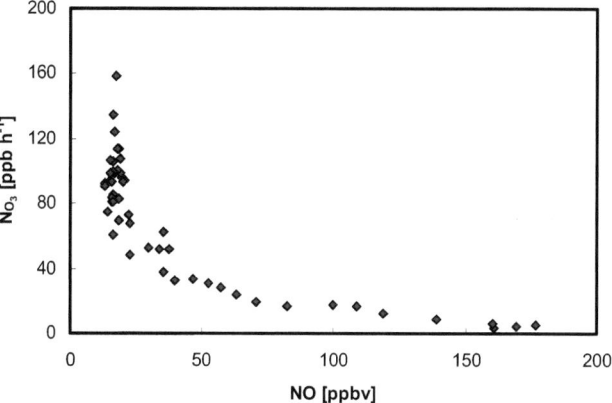

Fig. 3.3.10: Average diurnal profile of the net instantaneous production rate of O_3 (N_{O_3}) as a function of the NO concentration.

3.3.7 Photochemical Incremental Reactivity (PIR) Scale

It is important that policy actions undertaken to control the emissions of VOCs tackle those sources that contribute most to photochemical ozone formation (Derwent et al., 2007b and references therein). However, because of the lack of valid reactivity scales for ranking the different VOCs for the region of Santiago, a more reliable method that takes into consideration the kinetics and mechanisms in addition to the ambient concentrations has been used. The potential of each measured hydrocarbon to form ozone photochemically (scenario_3) has been determined based on the PIR scale defined earlier (see sec. 2.3.2.2). Table 3.3.1 shows the potentials of the measured VOCs expressed in terms of the $PIR_{adjusted}$. The $PIR_{adjusted}$ scales were calculated by multiplying the median value of the mixing ratio of each NMHC with the corresponding scale. The percentage contribution of the different VOCs can be calculated by dividing the value of the $PIR_{adjusted}$ scale of the different VOCs (VOC_i) with the total sum of $PIR_{adjusted}$ scales as following:

E 3.16 Contribution of VOC_i to ozone [%] $= \dfrac{PIR_{adjusted}(VOC_i)}{\sum PIR_{adjusted}(VOC_i)} \times 100$

Table 3.3.1: Ozone forming potentials of the measured VOCs based on the PIR$_{adjusted}$ scale. The ten most important species are marked in bold.

compound name	PIR $\Delta O_3/\Delta VOC$	PIR$_{adjusted}$**	mixing ratio (ppbv)		
			Average	Median	Max
propene	0.25	**0.39**	3.80	1.56	38.8
1,3-butadiene	1.05	0.13	0.15	0.12	0.41
trans-2-butene	2.13	0.23	0.18	0.11	0.86
cis-2-butene	1.64	0.16	0.16	0.10	0.67
3-methyl-1-butene	0.72	0.10	0.19	0.14	1.22
1-pentene	0.70	0.13	0.30	0.18	1.77
isoprene	2.85	**1.47**	0.67	0.51	1.84
trans-2-pentene	2.13	**0.30**	0.24	0.14	1.41
cis-2-pentene	2.77	0.27	0.15	0.10	0.74
2-methyl-2-buten	2.97	**0.58**	0.33	0.20	2.10
1-hexene	0.94	0.13	0.20	0.14	0.80
2,3-dimethyl-2-butene	6.56	**0.39**	0.10	0.06	0.52
α-pinene	2.15	**0.57**	0.41	0.27	1.95
propane	0.01	0.07	41.8	11.5	475
i-butane	0.03	0.05	2.95	1.57	18.0
n-butane	0.04	0.10	3.89	2.32	18.3
i-pentane	0.08	**0.32**	5.75	4.08	27.6
3-methylpentane	0.10	0.09	1.40	0.92	6.09
2-methylhexane	0.10	0.02	0.35	0.22	1.71
n-heptane	0.06	0.03	0.55	0.42	2.50
n-octane	0.06	0.01	0.34	0.23	1.82
benzene	0.02	0.03	2.13	1.43	9.22
toluene	0.18	**0.74**	6.30	4.11	32.7
n-decane	0.07	0.03	0.60	0.42	2.94
ethylbenzene	0.25	0.29	1.38	1.14	6.06
o-xylene	0.60	**0.91**	1.81	1.50	7.72
n-propylbenzene	0.20	0.05	0.36	0.26	1.68
1,3,5-trimethylbenzene	2.35	**0.90**	0.58	0.38	2.79
4-ethyltoluene	0.47	0.10	0.30	0.21	1.40
1,2,3-trimethylbenzene	1.44	0.24	0.27	0.17	1.31
styrene	0.46	0.07	0.22	0.16	1.02

**PIR$_{adjusted}$ scales were calculated by multiplying the median value of the mixing ratio of each VOC with the corresponding scale.

The ten most important VOCs contributing to ozone formation according to the PIR$_{adjusted}$ scale are (in order of importance): isoprene, o-xylene, 1,3,5-trimethylbenzene, toluene, 2-methyl-2-butene, α-pinene, 2,3-dimethyl-2-butene, propene, and i-pentane, trans-2-pentene (Table 1, column 3: PIR$_{adjusted}$, in bold). Anthropogenic VOC emissions represented by all aromatics, all alkenes (except isoprene and α-pinene) and all alkanes contribute most to ozone formation by ~77 %, followed by the biogenic hydrocarbons, 23 % represented by isoprene and α-pinene. In terms of an ozone source apportionment (scenario_4), total VOCs (excluding measured HCHO) was found to account for ~90 % of the total photochemical ozone formation, followed by HCHO, ~7 %, CO, ~1 % and methane, ~0.7 %.

3.3.8 Ozone Diurnal Structure

In order to elucidate the ozone diurnal structure, the diurnal variations of the simulated ozone concentration profiles produced by each of the measured hydrocarbons that have been identified as contributing most to ozone formation was investigated (scenario_4). Morning ozone formation is mainly driven by propene, almost exclusively up to ~10 h and with the only contribution peak at ~13 h (see Fig. 3.3.11). The highest contributions during daytime at ~15 h result from 1,3,5-trimethylbenzene and toluene followed by o-xylene and 2-methyl-2-butene (Fig. 3.3.12). The diurnal variations of ozone formation due to the emissions of isoprene and α-pinene in addition to ethylbenzene reach their maxima at ~17 h (Fig. 3.3.11) and hence are identified as possible sources of the observed summertime afternoon ozone shoulder.

However, because the ozone shoulder could not be simulated in the modelled ozone profile, other factors may contribute to the ozone shoulder. Unmeasured biogenic VOCs (e.g. terpenes and sesquiterpenes) may be an important factor that could contribute significantly to the afternoon shoulder. Meteorological conditions also play an important role because the dominant wind direction during daytime in Santiago is southwest with increasing wind speed during daytime (Gramsch et al., 2006) carrying biogenic emissions to the city centre. Thus, more speciated measurements of the biogenic VOCs and meteorological parameters combined with a photochemical trajectory model analysis are required to further elucidate the afternoon shoulder in the ozone diurnal profile.

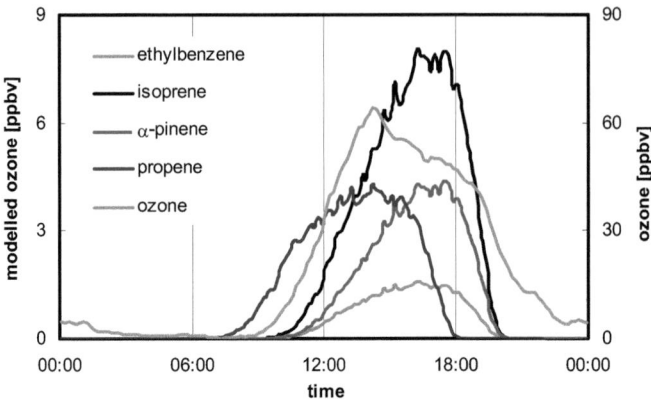

Fig. 3.3.11: Simulated average diurnal profiles of the contribution of different hydrocarbons to ozone formation during the time of the ozone shoulder and early morning.

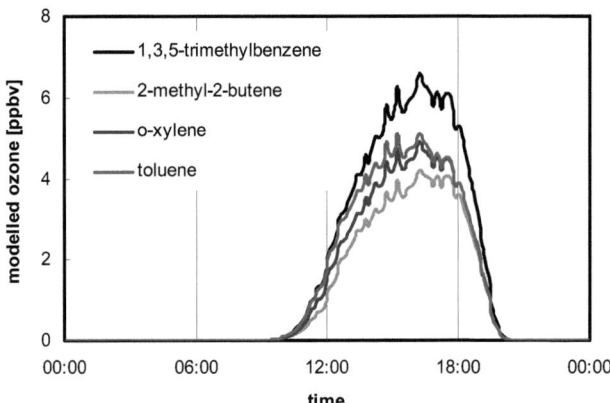

Fig. 3.3.12: Simulated average diurnal profiles of the contribution of different hydrocarbons to ozone formation during the daytime.

The afternoon ozone shoulder has previously been explained by the photooxidation of isoprene, based on the observation that peak isoprene concentrations coincide with the ozone shoulder (Rappenglück et al., 2005). However, the present study shows that although isoprene and α-pinene have a similar diurnal contribution to ozone formation (see Fig. 3.3.11), they have an apparent different average diurnal variation of their concentrations, as shown in Fig.

3.3.13. On the other hand, comparing propene with 2-methyl-2-butene shows that although both compounds have similar concentration profiles during daytime (Fig. 3.3.14), propene has a different diurnal contribution to ozone compared to 2-methyl-2-butene (Fig. 3.3.11 and Fig. 3.3.12). Thus, there is not necessarily a link between the contribution of a specific hydrocarbon to ozone formation and its corresponding diurnal profile of this hydrocarbon. That is because hydrocarbons produce ozone through an array of complex non-linear photochemical processes which may result in delayed ozone formation (Derwent et al. 2005).

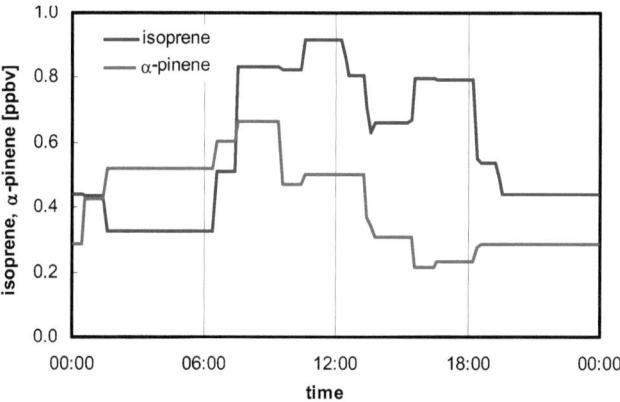

Fig. 3.3.13: Average measured diurnal variation of a) isoprene and α-pinene.

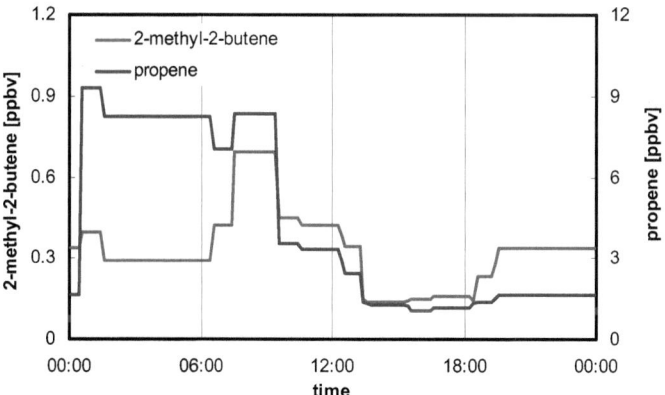

Fig. 3.3.14: Average measured diurnal variation of propene and 2-methly-2-butene.

3.3.9 Nitrous Acid Contribution to Ozone Formation

Ozone is a photochemical product and hence is not a significant radical source during early morning when solar radiation and ozone concentration are low. In Santiago, initiation of the radical chemistry and thus ozone formation in the early morning is mainly driven by the photolysis of HONO which contributes on average ~55 % and ~84 % to radical initiation during summer and winter, respectively (see sec. 3.2.6). Nevertheless, the role of HONO to ozone formation is a controversial issue. At high NO_x levels, Harris et al. (1982) have shown that increasing HONO levels have a marked effect on increasing the ozone formation yield. Similarly, Zhang et al. (2007) found that HONO concentrations of ~10 ppbv during early morning (based on the heterogeneous formation during the night) in the Hong Kong area may significantly enhance O_3 production during the day. In contrast, by using the community multiscale air quality modelling (CMAQ) Sarwar et al. (2008) have shown that adding HONO that results from heterogeneous reactions at night and surface photolysis during the day increases ozone levels by only ~4 % (or ~3 ppbv). However, in their study HONO was under predicted by the model and thus, the actual contribution of HONO to ozone formation was underestimated.

In the present study, the contribution of HONO to ozone formation has been estimated by comparing results from the HONO constrained base model to that unconstrained to the measured HONO (scenario_5). In scenario_5 the HONO concentration is thus only given by the PSS (E 3.9) caused by its gas phase chemistry. The percentage reduction in the modelled ozone concentration due to omitting the measured HONO (Fig. 3.3.15) compared to the base model scenario reached a morning maximum of 83 % at ~10 h and then decreased gradually to about 34 % at ~16 h. The average contribution of HONO to ozone formation during daytime (8 – 19 h) is ~37 %. This finding is in accordance with the fact that the OH radical is the driving force of the oxidation capacity of the atmosphere and with the high HONO contribution to the OH radical initiation rate in Santiago (see secs. 3.2.1 and 3.2.6). This result stresses the need to include measured HONO in ozone predicting models.

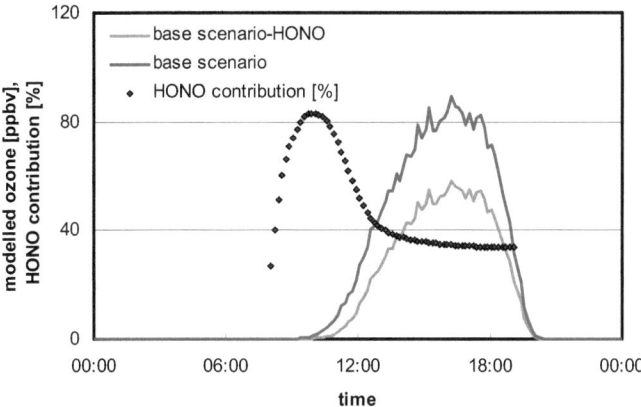

Fig. 3.3.15: Simulated ozone profiles of the base scenario and scenario_5, which was not constrained to measured HONO. In addition, the relative contribution of HONO to ozone formation is shown.

3.3.10 Impacts on the Ozone Control Strategy

The results of the present study show that summertime ozone photochemical formation in the city of Santiago is VOC-sensitive. The reduction of VOC concentrations by 50 % leads to a reduction in the O_3 peak by 36 %. However, the reduction in NO_x by 50 % leads to an increase in the ozone peak by 66 % (scenario_2, sec. 2.3.2.2). The reduction of both, VOCs and NO_x by 50 % (scenario_6) causes an increase in the peak ozone concentration by 10 %. This result shows that the reduction of VOCs in the urban area of Santiago is most appropriate to reduce ozone formation. However, it should be stressed that the results presented above are only representative for the city centre of Santiago and that the highly VOC limited condition observed in the present study are expected to change in the out-flow of the air mass from the city. Under these conditions, reduction in both, VOC and NO_x may be of higher importance. In a comparable study, Huang et al. (2001) also found that the reduction of VOCs by 50 % will lead to a reduction in peak ozone in the city of Tokyo by ~25 %, while a reduction of both NO_x and VOCs by 50 % lead to a reduction of peak ozone by only ~7 %.

The photochemical incremental reactivity scale, PIR (see sec. 3.3.7) can be used as a guide to reduce hydrocarbon emissions that contributes most to ozone formation. The measured biogenic hydrocarbons isoprene and α-pinene contribute 23 % to ozone formation, with isoprene calculated to have the largest ozone forming potential according to the PIR_{adjust} scale. However, it is obviously not possible to control biogenic emissions. Measured anthropogenic VOCs dominate the simulated ozone formation in Santiago (see sec. 3.3.7) and are mainly emitted from gasoline exhaust and evaporative emissions (Jorquera et al., 2004). Thus, reducing traffic emissions and gasoline vapour leaks from pump stations may significantly contribute to a decrease of ozone formation. Previous investigations showed that VOC emission reductions by using catalyst-equipped vehicles could reach 97-99 % compared to the emissions of non-catalyst cars (Stemmler et al., 2005). Thus, the further extended use of catalyst-equipped vehicles in Santiago may lead to a significant reduction in VOC emissions.

Future policy strategies of the environment commission of the metropolitan area of Santiago (CONAMA) include the reduction of NO_x emissions from stationary sources by 50 % by May 2010. In addition, the CONAMA authority of Santiago has implemented a major restructuring of the public transportation sector in February 2007 in order to reduce NO_x and PM_{10} emissions from mobile sources (Schreifel, 2008). However, based on the results of the current study, any ozone reduction policy should also consider reducing the VOC emissions in addition to NO_x. Comparably, it should be mentioned that much of the progress made so far with the control of photochemical ozone in Europe has been achieved through efficient control of VOC emissions from road transport and fuel evaporation (Derwent et al., 2007b).

Because the results of the current study are based on an urban scale box model, the aforementioned strategy should be considered only on a local scale. This is because different meteorological and geographical conditions (including seasonal change) may cause a VOC-sensitive regime to switch to NO_x sensitive (Sillman, 1995). Thus, additional seasonal investigations of the ozone formation and control, in other regions of Santiago, will be more representative in determining the controlling factors on a regional basis, especially when combined with photochemical analysis of air mass trajectories.

3.3.11 Conclusion

Photochemical ozone formation in the urban area of Santiago, Chile has been investigated using a detailed photochemical box model based on the MCMv3.1. The results of the model simulations have been compared with a set of potential empirical indicator relationships.

The results of the ozone model sensitivity and the empirical indicator relationship analyses show that photochemical ozone formation in Santiago is VOC limited. The model average net instantaneous ozone production rate (N_{O_3}) reached a maximum of ~160 ppbv h^{-1} at ~15 h and was associated with low ozone production efficiency (OPE) of ~1.5, which is a result of a highly polluted environment subjected to high NO_x levels. The ozone formation potentials of the different measured VOCs have been determined using the PIR scale. The high ozone formation potential of isoprene and α-pinene suggests that other biogenic emissions can have a significant effect on summertime photochemical ozone formation in the urban centre of Santiago. The diurnal ozone profile in Santiago has been investigated to determine the individual contributions of different hydrocarbons to ozone formation. Ozone formation in the morning was mainly driven by propene oxidation while the highest contributions during daytime result from 1,3,5-trimethylbenzene and toluene followed by o-xylene and 2-methyl-2-butene. The photooxidation of the biogenic species isoprene and α-pinene, were found to be possible contributors to the Santiago's summertime afternoon ozone shoulder. However, because the ozone shoulder was not reproduced in the simulated ozone profile, other factors may play an important role in its formation. Investigating the emissions of other unmeasured biogenic hydrocarbons may help elucidating the formation of the afternoon shoulder.

HONO was found to account on average for ~37 % of the simulated ozone formation. This result stresses the absolute need to include measured HONO levels in ozone predicting models.

On a local scale, the reduction of VOCs was found to be the most effective way in reducing ozone formation in the city centre of Santiago de Chile. Thus, the design of air quality control measures should also consider reducing VOC emissions in addition to NO_x. For the determination of the ozone control strategy on a regional scale, the analysis of ozone formation using a photochemical trajectory model is required.

4 Summary

The present study provides a detailed analysis of the tropospheric photochemical oxidation processes and ozone photochemical formation under typical polluted urban conditions taking the city of Santiago de Chile as an example. Two field measurement campaigns were carried out in the city of Santiago de Chile during summer and winter, 2005 namely March 8 - 20 and May 25 - June 07, 2005, respectively. The measured species included HONO, HCHO, O_3, NO_x, PAN, VOCs, CO, CO_2, $j(O^1D)$, $j(NO_2)$ and meteorological parameters.

Average mixing ratios of HONO, CO, NO and NO_2 in winter were higher than their corresponding values during summer. However, average and maximum mixing ratios of the photo-oxidation products O_3, PAN and HCHO (partially photochemically formed) during summer were higher than their corresponding values during winter owing to the higher summertime photochemical activity. The higher mixing ratios of the emitted species (CO, NO, NO_2) observed during winter are due to the higher stability of the lower boundary layer during winter. During both, summer and winter campaign, daytime HONO concentrations were significantly higher than in other polluted urban areas with the average HONO mixing ratio in winter higher than in summer. The high mixing ratios of HONO and the daytime maximum of the HONO/NO_x ratio in Santiago point to a very strong daytime HONO source. This is also confirmed by the lower daytime HONO/NO_x ratio during winter (lower photochemical activity) in comparison to that during summer.

The oxidation capacity of the atmosphere over the urban area of Santiago, Chile and its seasonal dependency has been studied during the two measurement campaigns. The measurement data including meteorological parameters were used to constrain a simple photostationary-state (PSS) model and a zero dimensional photochemical box model based on the MCMv3.1.

Total production/destruction rates of the OH and HO_2 in winter are ~2 times lower than those during summer, which is mainly due to the high NO_x levels and the lower photochemical activity during winter. The argument of the high NO_x conditions was further evidenced through an explicit analysis of the O_3-VOC-NO_x sensitivity analysis, performed during the summer. During both, winter and summer, there was a balance between the secondary production, $P_{OH}(HO_2 \rightarrow OH)$ and destruction, $L_{OH}(OH + VOC)$ rates of OH. This balance was also found to be fulfilled for other investigated urban studies. Therefore, initiation sources of HO_2 and RO_2 should not be considered as net OH sources under urban conditions. The excellent agreement between the OH concentration profiles evaluated by the MCM and PSS models during both, summer and winter shows that the major OH radical sources and sinks are incorporated in the PSS model and that the secondary sinks, $L_{OH}(OH + VOC)$ are balanced with the secondary sources $P_{OH}(HO_2 \rightarrow OH)$. For daytime conditions during summer and winter, HONO photolysis has the highest contribution of (55 %, 84 %) followed by alkenes ozonolysis (24 %, 9 %), HCHO photolysis (16 %, 6.5 %) and ozone photolysis (5 %, 0.5 %) to the OH initiation rate.

Summertime photochemical ozone formation in the urban area of Santiago, Chile has also been investigated using MCMv3.1. The results of the model simulations have been compared with a set of potential empirical indicator relationships. The ozone model sensitivity and the empirical indicator relationship analyses showed that photochemical ozone formation in Santiago is VOC limited. The ozone formation potentials of the different measured VOCs have been determined using the PIR scale. The high ozone formation potential of isoprene and α-pinene suggests that other biogenic emissions can have a significant effect on summertime photochemical ozone formation in the urban centre of Santiago. Investigating the emissions of other unmeasured biogenic hydrocarbons may help elucidating the formation of the ozone afternoon shoulder.

HONO was found to account on average for ~37 % of the simulated ozone formation. This result stresses the absolute need to include measured HONO levels in ozone predicting models.

On a local scale, the reduction of VOCs was found to be the most effective way in reducing ozone formation in the city centre of Santiago de Chile. Thus, the design of air quality control measures should also consider reducing VOC emissions in addition to NO_x.

5 Outlook

The current study presents the first detailed analysis of the oxidation capacity and ozone photochemical formation in the urban area of Santiago de Chile. As a result of this explicit study some questions have been raised, which can not be answered based on the available data.

5.1 Radical Budgets:

This study presented a detailed analysis of the radical budgets in Santiago de Chile, using the most explicit chemical model to date, MCMv3.1. However, the current study lacks of radical measurements, which would help to test the model performance. In addition, the investigation of differences that may result between the radical measurements and those simulated by the model would also lead to a better understanding of the photochemical oxidation processes in Santiago and other similar polluted urban environments. Furthermore, precise radical measurements would also help to better estimate the net HONO contribution to the radical budget. Therefore, another field campaign including radical measurements of OH, HO_2, RO_2 and NO_3 in addition to the above measured parameters, may help to better understand the oxidation capacity in polluted areas, especially the role of HONO photolysis and the oxidation of biogenic hydrocarbons.

In addition to the missing radical measurements, experimental determination of the OH radical reactivity would help to quantify unmeasured OH radical sinks, like unknown biogenic VOCs. The quantification of the total sinks of the OH radical will lead to a better understanding of the propagation processes during the photochemical degradation of VOCs in the atmosphere and would also help to verify atmospheric models.

Finally, for the quantification of radical budgets, the Master Chemical Mechanism code lacks for explicit mechanisms of the ozonolysis of cycloalkenes, which is the second most important alkenes category after internal alkenes in Santiago. Therefore, reactions schemes for these reactions needs to be developed and to be incorporated into the MCM,

5.2   Radical Balance:

A very important result from the current study is the balance obtained between secondary radical sources and sinks in urban environments, which was observed here for the first time. In the light of the current understanding of atmospheric chemistry given by the MCMv3.1, this balance obtained under different urban condition using different constraints under different meteorological conditions implies that the radical propagation process (i.e., given by the secondary production, $P_{OH}(HO_2 \rightarrow OH)$ and destruction $L_{OH}(OH + VOC)$ rates of OH) is not a net source of radicals. More focus on the role of hydrocarbons, especially biogenic hydrocarbons and their recycling processes is needed to shed more light on this balance ratio and its impact on the current understanding of atmospheric chemistry. Further investigation of the balance between secondary radical production and destruction in rural and remote areas are also needed for a better understanding of the photochemical oxidation processes under more clean conditions.

5.3   Ozone Control Strategy:

The analysis of the ozone photochemical formation and sensitivity was performed using a zero-dimensional photochemical box model and hence assumes that the box is continuously well mixed. Thus, the model will have some limitations for modelling of long-lived species (i.e., ozone). First, the amount of ozone entrained from aloft in the morning when the planetary boundary layer height start to increase is neglected. This may lead to an underestimation in the photochemically formed ozone. Therefore, a two layer (i.e., a residual layer, and nocturnal boundary layer) box model would be more suitable for the simulation of ozone and other long lived species. Second, since the applied model is a box model, drawn conclusions are predominantly limited to the local rather than regional scale. For the determination of the ozone control strategy on a regional scale, the analysis of ozone formation using a photochemical trajectory model is required.

The investigation of the ozone photochemical formation, particularly in the afternoon revealed that the emissions of unmeasured biogenic hydrocarbons might play an important role in the formation of the Santiago ozone afternoon shoulder. Therefore, more speciated measurements of biogenic hydrocarbons are required for better understanding and simulation of the ozone photochemical formation in Santiago de Chile.

In conclusion, the following work should be performed in order to reach a better understanding of the air pollution in the city of Santiago de Chile and other polluted urban areas:

Parallel radical and OH reactivity measurements.

More speciated measurements of biogenic hydrocarbons.

MCM updating with schemes of cycloalkene reactions with ozone (when available).

Trajectory model analysis of ozone photochemical formation.

6 Appendices

6.1 Appendix A: Measurement site

Coastal range   urban area of   Andes ranges
                 Santiago

Fig. 6.1.1: Satellite image of the metropolitan area of Santiago de Chile (LANSAT, 1998). Adopted from the Conama RM report "Evaluation of air quality in Santiago, 1997-2003".

Appendix

Fig. 6.1.2: USACH Measurements site at the University of Santiago de Chile (left). Measuremnts took place at the third floor of the physics department building (right).

6.2 Appendix B: VOC sampling system and calibration data

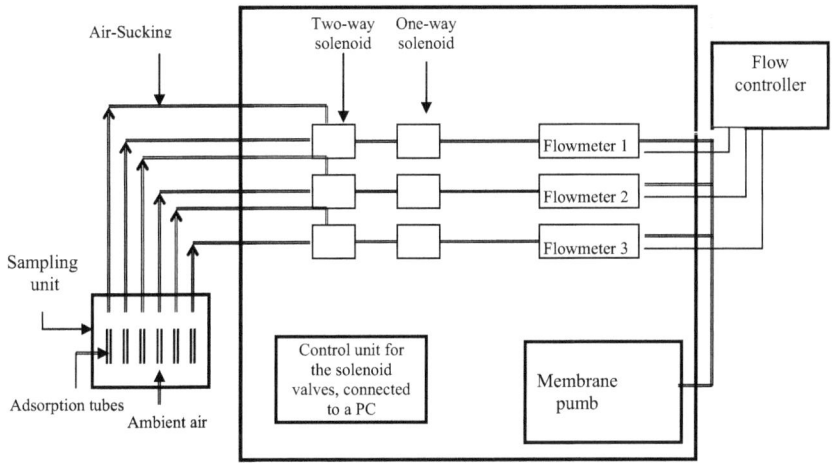

Fig. 6.2.1: Schematic diagram showing the sampling unit for VOCs.

Table 6.2.1: Standard gas mixture delivered by the National Physical Laboratory (NPL).

compound	molecular weight g/mole	mixing ratio ppb	uncertainty ppb	relative error $\Delta x/x$
ethene	28.05	22.03	0.44	0.02
ethyne	26.04	46.86	0.94	0.02
ethane	30.07	42.18	0.84	0.02
propene	42.08	21.19	0.42	0.02
propane	44.10	18.69	0.37	0.02
propyne	40.07	13.05	0.26	0.02
i-butane	58.12	6.22	0.12	0.02
1-butene	56.11	20.73	0.41	0.02
i-butene	56.11	20.86	0.42	0.02
butadien	54.09	28.15	0.56	0.02
n-butane	58.12	9.72	0.19	0.02
trans-2-butene	56.11	7.03	0.14	0.02
cis-2-butene	56.11	13.90	0.28	0.02
i-pentane	72.15	7.19	0.14	0.02
n-pentane	72.15	5.75	0.12	0.02
isoprene	68.12	15.12	0.30	0.02
trans-2-pentene	70.14	24.13	0.48	0.02
cis-2-pentene	70.14	12.42	0.25	0.02
2-methylpentane	86.18	9.24	0.18	0.02
3-methylpentane	86.18	14.32	0.29	0.02
n-hexane	86.18	16.26	0.33	0.02
benzene	78.11	27.85	0.56	0.02
cyclohexane	84.16	24.61	0.49	0.02
n-heptane	100.21	12.42	0.25	0.02
toluene	92.14	21.46	0.43	0.02
ethylbenzene	106.17	8.81	0.18	0.02
m-xylene	106.17	10.92	0.22	0.02
o-xylene	106.17	7.20	0.14	0.02
1,3,5-trimethylbenzene	120.20	6.47	0.13	0.02
1,2,4-trimethylbenzene	120.20	5.51	0.11	0.02

Table 6.2.2: List of the response factors (RF) calculated for the NPL standard mixture.

compound	RF [area/g]	SD ± [area/g]	relative error [%]
propene	1.66E+10	3.74E+08	2.25
propane	1.77E+10	5.54E+07	0.31
propyne	1.38E+10	1.46E+09	10.6
i-butane	1.69E+10	2.33E+08	1.38
1-butene, i-butene	1.64E+10	1.99E+08	1.21
butadien	1.56E+10	6.21E+08	3.97
n-butane	1.86E+10	2.39E+08	1.29
trans-2-butene	1.61E+10	4.68E+08	2.91
cis-2-butene	1.64E+10	2.40E+08	1.46
i-pentane	1.68E+10	2.51E+08	1.49
n-pentane	1.57E+10	1.72E+08	1.09
isoprene	1.50E+10	2.90E+08	1.93
trans-2-pentene	1.60E+10	3.16E+08	1.97
cis-2-pentene	1.55E+10	2.28E+08	1.47
2-methylpentane	1.66E+10	1.52E+08	0.92
3-methylpentane	1.67E+10	1.49E+08	0.89
n-hexane	1.61E+10	1.89E+08	1.17
benzene	1.72E+10	1.64E+08	0.95
cyclohexane	1.70E+10	8.32E+07	0.49
n-heptane	1.63E+10	1.51E+08	0.93
toluene	1.76E+10	5.43E+08	3.09
ethylbenzene	1.65E+10	8.82E+07	0.54
m-xylene	1.63E+10	1.22E+08	0.75
o-xylene	1.59E+10	4.11E+08	2.59
1,3,5-trimethylbenzene	1.52E+10	6.24E+08	4.10
1,2,4-trimethylbenzene	1.45E+10	5.01E+08	3.45

Table 6.2.3: Hydrocarbons mass concentrations in the RM calibration mixture.

compound	RT min	mass g/100 ml	SD ± g
propene	15.1	6.60E-09	2.40E-10
propane	15.9	8.41E-09	7.68E-11
i-butane (2-methylpropane)	22.7	1.07E-08	7.36E-11
1-butene, i-butene	25.1	9.08E-09	8.47E-11
n-butane	26.0	1.01E-08	1.09E-10
trans-2-butene	26.9	8.15E-09	1.26E-10
cis-2-butene	28.3	8.15E-09	9.34E-11
i-pentane (2-methylbutane)	32.1	7.73E-09	6.17E-11
1-pentene	33.7	9.50E-10	9.07E-12
n-pentane, 2-methyl-1-butene	34.0	1.26E-08	1.00E-10
isoprene	34.3	1.74E-09	1.23E-11
trans-2-pentene	34.5	2.44E-09	2.12E-11
cis-2-pentene	35.2	1.10E-09	2.26E-11
2,2-dimethylbutane	36.6	1.65E-08	1.48E-10
cyclopentene	37.7	2.24E-09	1.73E-11
methyl-tert-butyl ether, 2,3-dimethylbutane, cyclopentane	38.6	3.60E-08	2.94E-10
2-methylpentane	38.8	1.35E-08	1.28E-10
3-methylpentane	39.7	1.34E-08	1.01E-10
1-hexene	40.1	5.05E-09	3.47E-11
n-hexane, 2-ethyl-1-butene	40.7	2.52E-08	1.99E-10
2,3-dimethyl-1,3-butadiene	41.7	7.18E-09	7.61E-11
methylcyclopentane, 1-methyl-1-cyclopentene	42.7	2.21E-08	1.98E-10
2,3-dimethyl-2-butene	43.8	7.57E-09	1.95E-10
benzene	44.0	1.36E-08	1.46E-10
cyclohexane, 2,3-dimethylpentane	44.8	4.24E-08	3.63E-10
2-methylhexane	45.0	2.32E-08	2.15E-10
cyclohexene	45.7	1.36E-08	1.25E-10
1-heptene	45.9	1.52E-08	1.20E-10
2,2,4-trimethylpentane	46.3	2.93E-08	2.90E-10
n-heptane	46.6	1.67E-08	1.46E-10
1,4-cyclohexadiene	46.9	1.07E-08	2.85E-10
2,3,4-trimethylpentane	49.7	2.81E-08	2.22E-10
toluene	50.1	9.75E-09	1.42E-09
2-methylheptane	50.2	3.04E-08	5.02E-09
3-methylheptane	50.3	2.79E-08	3.22E-10
4-methylheptane, 1-methyl-1-cyclohexene	50.6	4.72E-08	3.86E-10
1-octene	51.2	2.88E-08	2.83E-10
n-octane	51.9	2.42E-08	2.56E-10
ethylbenzene	54.9	1.47E-08	2.83E-10
m-, p-xylene	55.5	2.96E-08	3.71E-10
styrene	56.3	3.31E-08	7.49E-10
o-xylene	56.6	1.64E-08	9.32E-11
α–pinene	59.1	6.04E-08	5.55E-10
n-propylbenzene	59.3	1.81E-08	4.17E-10
4-ethyltoluene (p-ethyltoluene)	59.7	1.39E-08	4.50E-10
1,3,5-trimethylbenzene	59.9	2.00E-08	4.44E-10
n-decane	61.1	1.46E-08	2.34E-09
1,2,4-trimethylbenzene, tetr. butylbenzene	61.3	3.35E-08	1.33E-09
1,2,3-trimethylbenzene	62.7	1.51E-08	1.67E-09
1,2,3,4-tetramethylbenzene	63.2	1.08E-08	1.20E-09

Table 6.2.4: List of the audit accuracies for the measured hydrocarbons in the RM mixture.

compound	audit accuracy [%]			
	3600 ml		7480 ml	
	Ave-RF	Ind-RF	Ave-RF	Ind-RF
propene	21	-6	26	1
propane	25	-4	29	2
i-butane (2-methylpropane)	41	11	34	0
1-butene, i-Butene	-3	8	-22	-9
n-butane	19	4	18	2
trans-2-butene	16	10	15	9
cis-2-butene	22	8	20	6
i-pentane (2-methylbutane)	40	-3	37	-8
1-pentene	35	0	34	-2
n-pentane, 2-methyl-1-butene	41	26	25	4
isoprene	69	9	65	0
trans-2-pentene	41	1	32	-13
cis-2-pentene	35	0	26	-14
2,2-dimethylbutane	49	-11	55	2
cyclopentene	29	3	31	5
methyl-tert-butyl ether, 2,3-dimethylbutane, cyclopentane	47	16	44	11
2-methylpentane	30	-9	35	0
3-methylpentane	29	-7	34	1
1-hexene	24	1	30	9
n-hexane, 2-ethyl-1-butene	23	-6	32	6
2,3-dimethyl-1,3-butadiene	21	-5	27	3
methylcyclopentane, 1-methyl-1-cyclopentene	42	3	40	1
2,3-dimethyl-2-butene	31	-3	36	5
benzene	20	13	-1	-9
cyclohexane, 2,3-dimethylpentane	34	0	35	1
2-methylhexane	16	-8	25	4
cyclohexene	35	-5	44	9
1-heptene	11	1	20	11
2,2,4-trimethylpentane	13	-3	21	6
n-heptane	11	0	19	10
1,4-cyclohexadiene	33	-2	42	11
2,3,4-trimethylpentane	14	0	19	5
toluene	-6	-10	0	-4
2-methylheptane	27	3	31	8
3-methylheptane	13	-3	17	2
4-methylheptane, 1-methyl-1-cyclohexene	14	-2	20	5
1-octene	11	3	15	7
n-octane	11	1	15	6
ethylbenzene	11	3	13	5
m-, p-xylene	9	-3	12	2
styrene	6	-2	10	1
o-xylene	10	5	14	8
a-pinene	13	0	18	6
n-propylbenzene	5	6	10	11
4-ethyltoluene (p-ethyltoluene)	1	8	8	14
1,3,5-trimethylbenzene	5	2	8	5
n-decane	0	3	-4	-1
1,2,4-trimethylbenzene, tetr. butylbenzene	8	-1	10	1
1,2,3-trimethylbenzene	4	-1	5	1
1,2,3,4-tetramethylbenzene	6	0	6	0

6.3 Appendix C: VOC measurements

Table 6.3.1: List of the measured hydrocarbons during field measurements in Santiago.

compound	RT	Ind-RF (area/g)	σ_{RF}	D_L (ppbv)
propene	15.10	1.24E+10	3.34E+08	0.07
propane	15.68	1.19E+10	2.12E+08	0.07
X7/4a	16.31	1.35E+10	2.99E+09	0.08
X7/4b	17.60	1.35E+10	2.99E+09	0.08
X5/13a	19.79	1.35E+10	2.99E+09	0.07
X5/13b	20.56	1.35E+10	2.99E+09	0.07
X5/13c	20.76	1.35E+10	2.99E+09	0.07
X5/13d	21.15	1.35E+10	2.99E+09	0.07
i-butane (2-methylpropane)	22.67	1.09E+10	1.08E+08	0.06
X13/10a	23.00	1.35E+10	2.99E+09	0.06
1-butene, i-Butene	25.11	1.84E+10	1.06E+09	0.20
butadiene	25.50	1.35E+10	2.99E+09	0.06
n-butane	26.00	1.39E+10	9.04E+07	0.04
X14/9a	26.73	1.35E+10	2.99E+09	0.06
trans-2-butene	26.90	1.55E+10	9.59E+07	0.04
X9/203a	27.15	1.35E+10	2.99E+09	0.06
X203/12a	27.85	1.35E+10	2.99E+09	0.06
cis-2-butene	28.15	1.41E+10	8.75E+07	0.04
X12/18a	28.84	1.35E+10	2.99E+09	0.05
X12/18b	29.89	1.35E+10	2.99E+09	0.05
X12/18c, X12/18d	30.43	1.35E+10	2.99E+09	0.05
3-methyl-1-butene	30.70	1.35E+10	2.99E+09	0.05
X18/25a	30.90	1.35E+10	2.99E+09	0.05
X18/25b	31.45	1.35E+10	2.99E+09	0.05
X18/25c	31.73	1.35E+10	2.99E+09	0.05
i-pentane (2-methylbutane)	32.14	9.67E+09	6.21E+08	0.06
X25/19a	32.60	1.35E+10	2.99E+09	0.05
X25/19b	32.90	1.35E+10	2.99E+09	0.05
X25/19c	33.15	1.35E+10	2.99E+09	0.05
X25/19d	33.23	1.35E+10	2.99E+09	0.05
1-pentene	33.68	1.07E+10	1.29E+08	0.01
n-pentane, 2-methyl-1-butene	33.95	1.30E+10	1.89E+08	0.06
isoprene	34.30	5.72E+09	1.21E+08	0.02
trans-2-pentene	34.50	9.83E+09	3.46E+08	0.01
cis-2-pentene	35.15	1.07E+10	3.91E+08	<0.01
X22/43a	35.25	1.35E+10	2.99E+09	0.04
2-methyl-2-butene	35.40	1.35E+10	2.99E+09	0.04
X22/43c	35.56	1.35E+10	2.99E+09	0.04
X22/43d	36.06	1.35E+10	2.99E+09	0.04
X22/43e	36.25	1.35E+10	2.99E+09	0.04
2,2-dimethylbutane	36.56	7.60E+09	7.22E+07	0.09
X43/17a	36.86	1.35E+10	2.99E+09	0.05
X43/17b	36.76	1.35E+10	2.99E+09	0.05
X43/17c	37.16	1.35E+10	2.99E+09	0.05
X43/17d	37.35	1.35E+10	2.99E+09	0.05
X43/17e	37.59	1.35E+10	2.99E+09	0.05
cyclopentene	37.70	1.21E+10	1.22E+08	0.02
X17/123a	37.88	1.35E+10	2.99E+09	0.05
X17/123b	37.99	1.35E+10	2.99E+09	0.05
X17/123c	38.20	1.35E+10	2.99E+09	0.05

Table 6.3.1: List of the measured hydrocarbons in Santiago de Chile (**Continued**).

compound	RT	Ind-RF (area/g)	σ_{RF}	D_L (ppbv)
methyl-tert-butyl ether, 2,3-dimethylbutane, cyclopentane	38.55	1.04E+10	8.15E+07	0.16
2-methylpentane	38.77	1.07E+10	2.38E+08	0.06
X45/46a	39.00	1.35E+10	2.99E+09	0.04
X45/46b	39.19	1.35E+10	2.99E+09	0.04
3-methylpentane	39.73	1.10E+10	9.99E+07	0.05
1-hexene	40.05	1.27E+10	9.73E+07	0.02
X37/47a	40.20	1.35E+10	2.99E+09	0.04
X37/47b	40.50	1.35E+10	2.99E+09	0.04
n-hexane, 2-ethyl-1-butene	40.74	1.20E+10	1.04E+08	0.06
X204/205a	41.00	1.35E+10	2.99E+09	0.04
X204/205b	41.04	1.35E+10	2.99E+09	0.04
X204/205c	41.30	1.35E+10	2.99E+09	0.04
X204/205d	41.42	1.35E+10	2.99E+09	0.04
X204/205e	41.50	1.35E+10	2.99E+09	0.04
2,3-dimethyl-1,3-butadiene	41.65	1.24E+10	7.41E+07	0.02
X205/41a	41.84	1.35E+10	2.99E+09	0.04
X205/41b	41.90	1.35E+10	2.99E+09	0.04
X205/41c	42.20	1.35E+10	2.99E+09	0.04
methylcyclopentane, 1-methyl-1-cyclopentene	42.66	9.91E+09	6.14E+07	0.07
X206/207a	42.90	1.35E+10	2.99E+09	0.04
X206/207b	42.98	1.35E+10	2.99E+09	0.04
X206/207c	43.15	1.35E+10	2.99E+09	0.04
X206/207d	43.21	1.35E+10	2.99E+09	0.04
X206/207e	43.55	1.35E+10	2.99E+09	0.04
2,3-dimethyl-2-butene	43.83	1.11E+10	1.02E+08	0.02
benzene	44.09	1.52E+10	9.01E+08	0.08
X28/42a	44.20	1.35E+10	2.99E+09	0.04
X28/42b	44.36	1.35E+10	2.99E+09	0.04
cyclohexane, 2,3-dimethylpentane	44.77	1.09E+10	6.60E+07	0.07
2-methylhexane	44.98	1.29E+10	9.64E+07	0.04
X62/30a	45.30	1.35E+10	2.99E+09	0.04
cyclohexene	45.57	1.02E+10	1.36E+08	0.05
1-heptene	45.93	1.48E+10	9.64E+07	0.02
X209/69a, X209/69b	46.11	1.35E+10	2.99E+09	0.03
2,2,4-trimethylpentane	46.19	1.39E+10	1.02E+08	0.03
X69/61a	46.37	1.35E+10	2.99E+09	0.03
n-heptane	46.55	1.47E+10	1.00E+08	0.02
X61/210a	46.70	1.35E+10	2.99E+09	0.04
X61/210b	46.83	1.35E+10	2.99E+09	0.04
1,4-cyclohexadiene	47.00	1.08E+10	1.06E+08	0.04
X210/70a	47.14	1.35E+10	2.99E+09	0.04
X210/70b	47.33	1.35E+10	2.99E+09	0.04
X210/70c	47.43	1.35E+10	2.99E+09	0.04
X210/70d	47.48	1.35E+10	2.99E+09	0.04
X210/70e	47.83	1.35E+10	2.99E+09	0.04
X210/70f	48.05	1.35E+10	2.99E+09	0.04
X210/70g	48.19	1.35E+10	2.99E+09	0.04
X210/70h	48.32	1.35E+10	2.99E+09	0.04
X210/70i	48.60	1.35E+10	2.99E+09	0.04
X210/70j	48.80	1.35E+10	2.99E+09	0.04
X210/70k	48.90	1.35E+10	2.99E+09	0.04
X210/70L	49.02	1.35E+10	2.99E+09	0.04
X210/70m	49.16	1.35E+10	2.99E+09	0.04

Table 6.3.1: List of the measured hydrocarbons in Santiago de Chile **(Continued)**.

compound	RT	Ind-RF (area/g)	σ_{RF}	D_L (ppbv)
X210/70n	49.30	1.35E+10	2.99E+09	0.04
X210/70q	49.52	1.35E+10	2.99E+09	0.04
2,3,4-trimethylpentane	49.75	1.42E+10	6.46E+07	0.03
toluene	50.05	1.59E+10	9.13E+07	0.01
2-methylheptane	50.22	1.24E+10	8.77E+08	0.03
3-methylheptane	50.35	1.40E+10	5.34E+07	0.02
4-methylheptane, 1-methyl-1-cyclohexene	50.59	1.39E+10	6.75E+07	0.06
X211/212a	50.86	1.35E+10	2.99E+09	0.03
X211/212b	51.10	1.35E+10	2.99E+09	0.03
X211/212c	51.15	1.35E+10	2.99E+09	0.03
1-octene	51.31	1.51E+10	6.72E+07	0.03
X212/81a	51.47	1.35E+10	2.99E+09	0.03
X212/81b	51.70	1.35E+10	2.99E+09	0.03
n-octane	51.83	1.49E+10	5.87E+07	0.02
X81/64a	52.10	1.35E+10	2.99E+09	0.03
X81/64b	52.37	1.35E+10	2.99E+09	0.03
X81/64c	52.61	1.35E+10	2.99E+09	0.03
X81/64d	52.75	1.35E+10	2.99E+09	0.03
X81/64e	52.95	1.35E+10	2.99E+09	0.03
X81/64f	53.15	1.35E+10	2.99E+09	0.03
X81/64g	53.38	1.35E+10	2.99E+09	0.03
X81/64h	53.54	1.35E+10	2.99E+09	0.03
X81/64i	53.79	1.35E+10	2.99E+09	0.03
X81/64j	53.93	1.35E+10	2.99E+09	0.03
X81/64k	54.07	1.35E+10	2.99E+09	0.03
X81/64l	54.18	1.35E+10	2.99E+09	0.03
X81/64m	54.29	1.35E+10	2.99E+09	0.03
X81/64n	54.49	1.35E+10	2.99E+09	0.03
X81/64o	54.67	1.35E+10	2.99E+09	0.03
X81/64p	54.88	1.35E+10	2.99E+09	0.03
ethylbenzene	54.95	1.51E+10	1.07E+08	0.02
X64/66a	55.05	1.35E+10	2.99E+09	0.03
X64/66b	55.16	1.35E+10	2.99E+09	0.03
m-, p-xylene	55.25	1.47E+10	1.14E+08	0.04
X67/145a	55.66	1.35E+10	2.99E+09	0.03
X67/145b	55.92	1.35E+10	2.99E+09	0.03
X67/145c	56.16	1.35E+10	2.99E+09	0.03
styrene	56.27	1.51E+10	1.03E+08	0.03
X145/65a	56.39	1.35E+10	2.99E+09	0.03
o-xylene	56.55	1.55E+10	5.95E+07	0.04
X65/124a	56.61	1.35E+10	2.99E+09	0.03
X65/124b	56.70	1.35E+10	2.99E+09	0.03
X65/124c	56.80	1.35E+10	2.99E+09	0.03
X65/124d	56.98	1.35E+10	2.99E+09	0.03
X65/124e	57.37	1.35E+10	2.99E+09	0.03
X65/124f	57.46	1.35E+10	2.99E+09	0.03
X65/124g	57.63	1.35E+10	2.99E+09	0.03
X65/124h	57.80	1.35E+10	2.99E+09	0.03
X65/124i	58.00	1.35E+10	2.99E+09	0.03
X65/124j	58.36	1.35E+10	2.99E+09	0.03
X65/124k	58.44	1.35E+10	2.99E+09	0.03
X65/124l	58.63	1.35E+10	2.99E+09	0.03

Appendix

Table 6.3.1: List of the measured hydrocarbons in Santiago de Chile (**Continued**).

compound	RT	Ind-RF (area/g)	σ_{RF}	D_L (ppbv)
X65/124m	58.83	1.35E+10	2.99E+09	0.03
α-pinene	59.12	1.43E+10	6.60E+07	0.07
n-propylbenzene	59.34	1.67E+10	5.34E+07	0.02
X85/91a	59.47	1.35E+10	2.99E+09	0.03
X85/91b	59.60	1.35E+10	2.99E+09	0.03
4-ethyltoluene (p-ethyltoluene)	59.71	1.77E+10	7.86E+07	0.01
1,3,5-trimethylbenzene	59.88	1.61E+10	3.10E+07	0.03
X88/120a	60.14	1.35E+10	2.99E+09	0.03
X88/120b	60.40	1.35E+10	2.99E+09	0.03
X88/120c	60.56	1.35E+10	2.99E+09	0.03
X88/120d	60.84	1.35E+10	2.99E+09	0.03
n-decane	61.06	1.70E+10	8.76E+07	0.02
1,2,4-trimethylbenzene, tetr. butylbenzene	61.20	1.50E+10	7.39E+07	0.04
X213/86a	61.34	1.35E+10	2.99E+09	0.03
X213/86b	61.70	1.35E+10	2.99E+09	0.03
X213/86c	61.79	1.35E+10	2.99E+09	0.03
X213/86d	62.03	1.35E+10	2.99E+09	0.03
X213/86e	62.12	1.35E+10	2.99E+09	0.03
X213/86f	62.26	1.35E+10	2.99E+09	0.03
X213/86g	62.44	1.35E+10	2.99E+09	0.03
X213/86h	62.55	1.35E+10	2.99E+09	0.03
1,2,3-trimethylbenzene	62.69	1.58E+10	6.02E+07	0.01
X86/115a	62.80	1.35E+10	2.99E+09	0.03
X86/115b	62.90	1.35E+10	2.99E+09	0.03
1,2,3,4-tetramethylbenzene	63.10	1.56E+10	6.73E+07	0.02
X115a	63.41	1.35E+10	2.99E+09	0.02
X115b	63.58	1.35E+10	2.99E+09	0.02
X115c	63.76	1.35E+10	2.99E+09	0.02
X115d	63.94	1.35E+10	2.99E+09	0.02
X115e	64.07	1.35E+10	2.99E+09	0.02
X115f	64.20	1.35E+10	2.99E+09	0.02
X115g	64.45	1.35E+10	2.99E+09	0.02
X115h	64.52	1.35E+10	2.99E+09	0.02
X115i	64.75	1.35E+10	2.99E+09	0.02
X115j	65.08	1.35E+10	2.99E+09	0.02
X115k	65.27	1.35E+10	2.99E+09	0.02
X115l	65.53	1.35E+10	2.99E+09	0.02
X115m	65.74	1.35E+10	2.99E+09	0.02
X115n	65.97	1.35E+10	2.99E+09	0.02
X115o	66.03	1.35E+10	2.99E+09	0.02
X115p	66.35	1.35E+10	2.99E+09	0.02
X115q	66.46	1.35E+10	2.99E+09	0.02
X115r	66.81	1.35E+10	2.99E+09	0.02
X115s	67.05	1.35E+10	2.99E+09	0.02
X115t	67.38	1.35E+10	2.99E+09	0.02
X115u	67.60	1.35E+10	2.99E+09	0.02
X115w	67.86	1.35E+10	2.99E+09	0.02
X115x	68.10	1.35E+10	2.99E+09	0.02
X115y	68.30	1.35E+10	2.99E+09	0.02
X115z	68.58	1.35E+10	2.99E+09	0.02
X115aa	68.73	1.35E+10	2.99E+09	0.02
X115ab	68.97	1.35E+10	2.99E+09	0.02

Table 6.3.1: List of the measured hydrocarbons in Santiago de Chile (**Continued**).

compound	RT	Ind-RF (area/g)	σ_{RF}	D_L (ppbv)
X115ac	69.16	1.35E+10	2.99E+09	0.02
X115ad	69.30	1.35E+10	2.99E+09	0.02
X115ae	69.61	1.35E+10	2.99E+09	0.02
X115af	69.90	1.35E+10	2.99E+09	0.02
X115ag	70.05	1.35E+10	2.99E+09	0.02
X115ah	70.30	1.35E+10	2.99E+09	0.02
X115ai	70.58	1.35E+10	2.99E+09	0.02
X115aj	71.05	1.35E+10	2.99E+09	0.02
X115ak	71.19	1.35E+10	2.99E+09	0.02
X115al	71.34	1.35E+10	2.99E+09	0.02
X115am	71.70	1.35E+10	2.99E+09	0.02
X115an	71.90	1.35E+10	2.99E+09	0.02
X115ao	72.17	1.35E+10	2.99E+09	0.02
X115ap	72.53	1.35E+10	2.99E+09	0.02
X115aq	72.78	1.35E+10	2.99E+09	0.02
X115ar	78.17	1.35E+10	2.99E+09	0.02

7 Bibliography

Aceituno, P.: On the functioning of the southern oscillation in the South American sector: part I surface climate, *Monthly Weather Review*, **116**, 505-524, 1988.

Acker, K., Möller, D., Wieprecht, W., Meixner, F. X., Bohn, B., Gilge, S., Plass-Dülmer, C. and Berresheim, H.: Strong Daytime Production of OH from HNO_2 at a Rural Mountain Site, *Geophysical Research Letters*, **33**, L02809, doi: 10.1029/2005GL024643, 2006a.

Acker, K., Febo, A., Trick, S., Perrino, C., Bruno, P., Wiesen, P., Möller, D., Wieprecht, W., Auel, R., Guisto, M., Geyer, A., Platt, U. and Allegrini, I.: Nitrous Acid in the Urban Area of Rome, *Atmospheric Environment*, **40**, 3123-3133, 2006b.

Adonis, M. and Gil, L.: Polycyclic aromatic hydrocarbon levels and mutagenicity of inhalable particulate matter in Santiago, Chile, *Inhalation Toxicology*, **12**, 1173-1183, 2000.

Alicke, B., Platt, U., and Stutz, J.: Impact of nitrous acid photolysis on the total hydroxyl radical budget during the Limitation of Oxidant Production/Pianura Padana Produzione di Ozono study in Milan, *Journal of Geophysical Research*, **107** (D22), 8196, doi: 10.1029/2000JD000075, 2002.

Alicke, B., Geyer, A., Hofzumahaus, A., Holland, F., Konrad, S., Pätz, H. W., Schäfer, J., Stutz, J., Volz-Thomas, A. and Platt, U.: OH formation by HONO photolysis during the BERLIOZ experiment, *Journal of Geophysical Research*, **108** (D4), 8247, doi: 10.1029/2001JD000579, 2003.

Ammann, M., Rössler, E., Strekowski, R. and George, C.: Nitrogen Dioxide Multiphase Chemistry: Uptake Kinetics on Aqueous Solutions Containing Phenolic Compounds, *Physical Chemistry Chemical Physics*, **7**, 2513-2518, 2005.

Appel, B. R., Winer, A. M., Tokiwa, Y. and Biermann, H. W.: Comparison of Atmospheric Nitrous Acid Measurements by Annular Denuder and Optical Absorption Systems, *Atmospheric Environment*, **24 A**, 611-616, 1990.

Arens, F., Gutzwiller, L., Gäggeler, H. W. and Ammann, M.: The Reaction of NO_2 with Solid Anthrarobin (1,2,10-trihydroxy-anthracene), *Physical Chemistry Chemical Physics*, **4**, 3684-3690, 2002.

Atkinson, R., Aschmann, S. M., Arey, J., Shorees, B. J.: Formation of OH radicals in the gas phase reactions of O3 with a series of terpenes, *Journal of Geophysical Research*, **97** (D5), 6065-6073, doi:10.1029/2004GL021322, 1992.

Atkinson, R., Baulch, D. L., Cox, R. A., Crowley, J. N., Hampson, R. F., Hynes, R. G., Jenkin, M. E., Rossi, M. J., and Troe, J.: Evaluated kinetic and photochemical data for atmospheric chemistry: Volume I - gas phase reactions of O_x, HO_x, NO_x and SO_x species, *Atmospheric Chemistry and Physics*, **4**, 1461-1738, 2004.

Bejan, I., Abd El Aal, Y., Barnes, I., Benter, T., Bohn, B., Wiesen, P. and Kleffmann J.: The Photolysis of Ortho-Nitrophenols: A new Gas Phase Source of HONO, *Physical Chemistry Chemical Physics*, **8**, 2028–2035, 2006.

Bloss, C., Wagner, V., Bonzanini, A., Jenkin, M. E., Wirtz, K., Martin-Reiejo, M. and Pilling, M. J.: Evaluation of detailed aromatic mechanisms (MCMv3 and MCMv3.1) against environmental chamber data, *Atmospheric Chemistry and Physics*, **5**, 623-639, 2005a.

Bloss, C., Wagner, V., Jenkin, M. E., Volkamer, R., Bloss, W. J., Lee, J. D., Heard, D. E., Wirtz, K., Martin-Reviejo, M., Rea, G., Wenger, J. C. and Pilling, M. J.: Development of a detailed chemical mechanism (MCMv3.1) for the atmospheric oxidation of aromatic hydrocarbons, *Atmospheric Chemistry and Physics*, **5**, 641–664, 2005b.

Cakmak, S., Dales, R. E. and Blanco, V. C.: Air Pollution and Mortality in Chile: Susceptibility among the Elderly, *Environmental Health Perspectives*, **115**, 524-527, 2007.

Cardelino, C. and Chameides, W. L.: An observation-based model for analyzing ozone precursor relationships in the urban atmosphere, *Journal of Air Waste Management Association*, **45**, 161–180, 1995.

Cárdenas, L. M., Brassington, D. J., Allan, B. J., Coe, H., Alicke, B., Platt, U., Wilson, K. M., Plane, J. M. C. and Penkett, S. A.: Intercomparison of formaldehyde measurements in clean and polluted atmospheres, *Atmospheric Environment*, **37**, 53-80, 2000.

Carr, S., Heard, D. E., Blitz, M. A.: Comment on atmospheric hydroxyl radical production from electronically excited NO_2 and H_2O", *Science*, **324** (5925), 336b, doi: 10.1126/science.1166669, 2009.

Carslaw, D. C. and Beevers, S. D.: Estimation of road vehicle primary NO_2 exhaust emission fractions using monitoring data in London, *Atmospheric Environment*, **39**, 167-177, 2005.

Carter, W. P. L.: Development of ozone reactivity scales for volatile organic compounds, *Journal of Air and Waste Management Association*, **44**, 881-899, 1994.

Carvacho, O. F., Trzepla-Nabaglo, K., Ashbaugh, L. L., Flocchini, R. G., Melin, P., Celis, J.: Elemental composition of springtime aerosol in Chillan, Chile, *Atmospheric Emnvironment*, **38**, 5349-5352, 2004.

Chatani, S., Shimo, N., Matsunaga, S., Kajii, Y., Kato, S., Nakashima, Y., Miyazaki, K., Ishii, K., and Ueno, H.: Sensitivity analyses of OH missing sinks over Tokyo metropolitan area in the summer of 2007, *Atmospheric Chemistry and Physics Discussion*, **9**, 18479-18509, 2009.

Chen, T.-Y., Simpson, I. J., Blake, D. R. and Rowland, F. S.: Impact of the leakage of liquefied petroleum gas (LPG) on Santiago air quality, *Geophysical Research Letters*, **28**, 2193-2196, 2001.

Cheng, Y., Wang, X., Liu, Z., Bai, Y. and Li, J.: A new method for quantitatively characterizing atmospheric oxidation capacity, *Science in China Series B: Chemistry*, **51**, 1102-1109, 2008.

CONAMA RM: Evaluation of the air quality in Santiago de Chile, 1997-2003, CONAMA Regional Metropolitan area of Santiago, 2003, www.conamarm.cl.

Crowley, J. N. and Carl, S. A.: OH formation in the photoexcitation of NO_2 beyond the dissociation threshold in the presence of water vapor, *Journal of Physical Chemistry*, **101**, 4178–4184, 1997.

Curtis, A. R., Sweetenham, W. P.: FACSIMILE Release H User's Manual, *AERE Report R11771, HMSO, London*, 1987.

Dasgupta, P. K., Dong, S., Hwang, H., Yang, H.-C. and Genfa, Z.: Continuous liquid phase fluorometry coupled to a diffusion scrubber for the real-time determination of atmospheric formaldehyde and sulfur dioxide, *Atmospheric Environment*, **22**, 949-963, 1988.

Derwent, R. G., Jenkin, M. E. and Saunders, S. M.: Photochemical ozone creation potentials for a large number of reactive hydrocarbons under European conditions, *Atmospheric Environment*, **30**, 181–199, 1996.

Derwent, R. G., Jenkin, M. E., Saunders, S. M. and Pilling, M. J.: Photochemical ozone creation potentials for organic compounds in northwest Europe calculated with a Master Chemical Mechanism, *Atmospheric Environment*, **32**, 2429–2441, 1998.

Derwent, R. G; Jenkin, M. E; Saunders, S. M; Pilling and M. J; Passant, N. R. Multi-day ozone formation for alkenes and carbonyls investigated with a master chemical mechanism under European conditions, *Atmospheric Environment*, **39**, 627-635, 2005.

Derwent, R. G., Jenkin, M. E., Passant, N. R. and Pilling, M. J.: Reactivity-based strategies for photochemical ozone control in Europe, *Environmental Science & Policy*, **10**, 445-453, 2007a.

Derwent, R. G., Jenkin, M. E., Passant, N. R. and Pilling, M. J.: Photochemical ozone creation potentials (POCPs) for different emission sources of organic compounds under European condition estimated with a Master Chemical Mechanism, *Atmospheric Environment*, **41**, 2570-2579, 2007b.

Diaz, S., Camilion, C., Deferrari, G., Fuenzalida, H., Armstrong, R., Booth, C., Paladini, A., Cabrera, S., Casiccia, C., Lovengreeng, C., Pedroni, J., Rosales, A., Zagarese, H. and Vernet, M.: Ozone and UV Radiation over Southern South America: Climatology and Anomalies, *Photochemistry and Photobiology*, **82**, 834-843, 2006.

Di Carlo, P., Brune, W.H., Martinez, M., Harder, H., Lesher, R., Ren, X., Thornberry, T., Carroll, M.A., Young, V., Shepson, P.B., Riemer, D., Apel, E. and Campbell, C.: Missing OH reactivity in a forest: Evidence for unknown reactive biogenic VOCs, *Science*, **304**, 722–725, 2004.

Dong, S. and Dasgupta, P. K.: Solubility of gaseous formaldehyde in liquid water and generation of trace Standard gaseous formaldehyde, *Environmental Science & Technology*, 20, 637-640, 1986.

Dong S. and Dasgupta, P. K.: Fast flow injection analysis of formaldehyde in atmospheric water, *Environmental Science & Technology*, **21**, 581-588, 1987.

Dusanter, S., Vimal, D., Stevens, P. S., Volkamer, R., Molina, L. T., Baker, A., Meinardi, S.,. Blake, D. R., Sheehy, P., Merten, A., Zhang, R., Zheng, J.,. Fortner, E. C., Junkermann, W., Dubey, M. K., Rahn, T.,. Eichinger, W. E., Lewandowski, P., Prueger, J. and Holder, H.: Measurements of OH and HO_2 concentrations during the MCMA-2006 field campaign-Part 2: Model comparison and radical budget, *Atmospheric Chemistry and Physics*, **9**, 6655-6675, 2009.

Emmerson, K. M., Carslaw, N., Carpenter, L. J., Heard, D. E., Lee, J. D. and Pilling, M. J.: Urban atmospheric chemistry during the PUMA campaign 1: Comparison of modelled OH and HO_2 concentrations with measurements, *Journal of Atmospheric Chemistry*, **52**, 143–164, 2005a.

Emmerson, K. M., Carslaw, N. and Pilling, M. J.: Urban Atmospheric Chemistry during the PUMA Campaign. 2: Radical budgets for OH, HO_2 and RO_2, *Journal of Atmospheric Chemistry*, **52**, 165–183, 2005b.

Emmerson, K. M., Carslaw N., Carslaw D. C., Lee, J. D., McFiggans, G., Bloss, W. J., Gravestock, T., Heard, D. E., Hopkins, J., Ingham, T., Pilling, M. J., Smith, S. C., Jacob, M. and Monks, P. S.: Free radical modelling studies during the UK TORCH Campaign in summer 2003, *Atmospheric Chemistry and Physics*, **7**, 167–181, 2007.

Finlayson-Pitts, B. J. and Pitts, Jr., J. N.: Chemistry of the Upper and Lower Atmosphere, *Academic Press*, 2000.

Finlayson-Pitts, B. J., Wingen, L. M., Sumner, A. L., Syomin, D. and Ramazan, K. A.: The Heterogeneous Hydrolysis of NO_2 in Laboratory Systems and in Outdoor and Indoor Atmospheres: An Integrated Mechanism, *Physical Chemistry Chemical Physics*, **5**, 223-242, 2003.

Friedfeld, S., Fraser, M., Ensor, K., Tribble, S., Rehle, D., Leleux D. and Tittel, F.: Statistical analysis of primary and secondary atmospheric formaldehyde, *Atmospheric Environment*, **36**, 4767-4775, 2002.

Gallardo, L., Olivares, G., Langner, J. and Aarhus, B.: Coastal lows and sulfur air pollution in Central Chile, *Atmospheric Environment*, **36**, 3829-3841, 2002.

Garcia, A. R., Volkamer, R., Molina, L. T., Molina, M. J., Samuelson, J., Mellqvist, J., Galle, B., Herndon, S. C. and Kolb, C. E.: Separation of emitted and photochemical formaldehyde in Mexico City using a statistical analysis and a new pair of gas-phase tracers, *Atmospheric Chemistry and Physics*, **6**, 4545-4557, 2006.

García-Huidobro, T., Marshall, F. and Bell, J.: A risk assessment of potential crop losses due to ambient SO_2 in the central regions of Chile, *Atmospheric Environment*, **35**, 4903–4915, 2001.

Gear, C. W.: Numerical initial value problems in ordinary differential equations, *Prentice-Hall, inc., Engelwood Cliffs, New Jersey, US*, 1971.

George, L. A., Hard, T. M. and O'Brien, R. J.: Measurement of free radicals OH and HO_2 in Los Angeles smog, *Journal of Geophysical Research*, **104** (D9), 11,643–11,656, 1999.

George, C., Strekowski, R. S., Kleffmann, J., Stemmler, K. and Ammann, M.: Photoenhanced Uptake of Gaseous NO_2 on Solid Organic Compounds: A Photochemical Source of HONO?, *Faraday Discussion*, **130**, 195-210, 2005.

Geyer, A., Alicke, B., Konrad, S., Schmitz, T., Stutz, J. and Platt, U.: Chemistry and oxidation capacity of the nitrate radical in the continental boundary layer near Berlin, *Journal of Geophysical Research*, **106**, 8013-8025, 2001.

Gil, L., King, L. and Adonis, M.: Trends of polycyclic aromatic hydrocarbon levels and mutagenicity in Santiago's inhalable airborne particles in the period 1992-1996, *Inhalation Toxicology*, **12**, 1185-1204, 2000.

Goldstein, A. H. and Galbally, I. E.: Known and Unexplored Organic Constituents in the Earth's Atmosphere, Environmental Science and Technology, **41**, 1514–1521, 2007.

Gramsch, E., Catalan, L., Ormeno, I. and Palma, G.: Traffic and seasonal dependence of teh light absorption coefficient in santiago de Chile, Applied Optics, **39**, 4895-4901, 2000.

Gramsch, E., Cereceda-Balic F., Ormeno, I, Palma, G. and Oyola, P.: Use of the light absorption coefficient to monitor elemental carbon and $PM_{2.5}$. Example of Santiago de Chile, *Journal of the Air and Wast Management Association*, **54**, 799-808, 2004.

Gramsch, E., Cereceda-Balic F., Oyola, P. and von Baer, D.: Examination of pollution trends in Santiago de Chile with cluster analysis of PM10 and Ozone data, *Atmospheric Environment*, **40**, 5464-5475, 2006.

Gramsch, E., Gidhagen, L., Wahlin, P., Oyola, P. and Moreno, F.: Predominance of soot-mode ultrafine particles in Santiago de Chile: Possible sources, *Atmospheric Environment*, 43, 2260-2267, 2009.

Gustafsson, R. J., Orlov, A., Griffiths, P. T., Cox R. A., and Lambert, R. M.: Reduction of NO_2 to nitrous acid on illuminated titanium dioxide aerosol surfaces: implications for photocatalysis and atmospheric chemistry, *Chemical Communications*, 3936–3938, doi: 10.1039/b609005b, 2006.

Gutzwiller, L., Arens F., Baltensperger, U., Gäggeler H. W. and Ammann, M.: Significance of Semivolatile Diesel Exhaust Organics for Secondary HONO Formation, *Environmental Science and Technology*, 36, 677-682, 2002.

Haagen-Smit, A. J., Fox, M. M.: Photochemical ozone formation with hydrocarbons and automobile exhaust, *Journal of Air Pollution Control Association*, 4, 105-109, 1954.

Hak, C., Pundt, I., Trick, S., Kern, C., Platt, U., Dommen, J., Ord´o˜nez, C., Pr´ev^ot, A. S. H., Junkermann, W., Astorga-Llor´ens, C., Larsen, B. R., Mellqvist, J., Strandberg, A., Yu, Y., Galle, B., Kleffmann, J., L¨orzer, J. C., Braathen, G. O., and Volkamer, R.: Intercomparison of four different in-situ techniques for ambient formaldehyde measurements in urban air, *Atmospheric Chemistry and Physics*, 5, 2881-2900, 2005.

Harris, G. W., Carter, W. P. L., Winer A. M., Pitts J. N., Platt, U. and Perner, D.: Observations of nitrous acid in the Los Angeles atmosphere and implications for predictions of ozone-precursor relationships, *Environmental Science and Technology*, 16, 414-419, 1982.

Harrison, R. M., Yin, J., Tilling, R. M., Cai, X., Seakins, P.W., Hopkins, J. R., Lansley, D. L., Lewis, A. C., Hunter, M. C., Heard, D. E., Carpenter, L. J., Creasey, D. J., Lee, J. D., Pilling, M. J., Carslaw, N., Emmerson, K. M., Redington, A., Derwent, R. G., Ryall, D., Mills, G. and Penkett, S. A.: Measurement and modelling of air pollution and atmospheric chemistry in the U.K. west midlands conurbation: overview of the PUMA consortium project, *Science of Total Environment*, 360, 5-25, 2006.

Hayman, G. D.: Effects of Pollution Control on UV Exposure, AEA Technology Final Report (Ref: AEA/RCEC/22522001/R/002 Issue 1) AEA Technology, Oxfordshire, 1997.

Heard, D. E. and Pilling, M. J.: Measurement of OH and HO2 in the reoposphere, *Chemical Reviews*, **103**, 5163-5198, 2003.

Heard, D. E., Carpenter, L. J., Creasey, D. J., Hopkins, J. R., Lee, J. D., Lewis, A. C., Pilling, M. J., Seakins, P. W., Carslaw, N. and Emmerson K. M.: High levels of the hydroxyl radical in the winter urban troposphere, *Geophysical Research Letters*, **31**, L18112, 2004.

Heland, J., Kleffmann, J., Kurtenbach, R. and Wiesen, P.: A New Instrument to Measure Gaseous Nitrous Acid (HONO) in the Atmosphere, *Environmental Science and Technology*, **35**, 3207-3212, 2001.

Hofzumahaus, A., Rohrer, F., Lu, K., Bohn, B., Brauers, T., Chang, C. C., Fuchs, H., Holland, F., Kita, K., Kondo, Y., Li, X., Lou, S., Shao, M., Zeng, L., Wahner, A., and Zhang, Y.: Amplified Trace Gas Removal in the Troposphere, *Science*, **324** (5935), 1702–1704, 2009.

Holland, F., Hofzumahaus, A., Schäfer, J., Kraus, A. and Pätz, H.-W.: Measurements of OH and HO_2 Radical Concentration and Photolysis Frequencies during BERLIOZ; *Journal of Geophysical Research*, **180** (D4), 8246, doi: 10.1029/2001JD001393, 2003.

Horvath, H., and Trier, A.: A study of the aerosol of Santiago de Chile-I. Light extension coefficients, *Atmospheric Environment*, **27A**, 371-384, 1993.

Huang, H., Akustu, Y., Arai, M. and Tamura, M.: Analysis of photochemical pollution in summer and winter using a photochemical box model in the centre of Tokyo, Japan, *Chemosphere*, **44**, 223-230, 2001.

Huang, G., Zhou, X., Deng, G., Qiao, H. and Civarolo, K.: Measurements of atmospheric nitrous acid and nitric acid, *Atmospheric Environment*, **36**, 2225-2235, 2002.

Ilabaca, M., Olaeta, I., Campos, E., Villaire, J., Tellez-Rojo, M. and Romieu, I.: Association between levels of fine particulate and emergency visits for pneumonia and other respiratory illnesses among children in Santiago, Chile, *Journal of Air and Waste Management Association*, **49**, 154–163, 1999.

Jenkin, M. E., Saunders, S. M. and Pilling, M. J.: The tropospheric degradation of volatile organic compounds: a protocol for mechanism development, *Atmospheric Environment*, **31**, 81–104, 1997.

Jenkin, M. E., Saunders, S. M., Wagner, V. and Pilling, M. J.: Protocol for the development of the Master Chemical Mechanism, MCM V3 (Part B): tropospheric degradation of aromatic volatile organic compounds, *Atmospheric Chemistry and Physics*, **3**, 181-193, 2003.

Johnson, D. and Marston, G.: The gas-phase ozonolysis of unsaturated volatile organic compounds in the troposphere, *Chemical Society Reviews*, **37**, 699-716, 2008.

Jorquera, H., Perez, R., Cipriano, A., Espejo, A., Letelier, M. V. and Acuna, G.: Forecasting daily maximum levels at Santiago, Chile, *Atmospheric Environment*, **32**, 3415-3424, 1998a.

Jorquera H., Olguín, C., Ossa, F., Pérez, R., Solar, I., Encalada, O.: An assessment of photochemical pollution at Santiago, Chile. In: Brebbia, C.A., Ratto, C.F. and Power, H., Editors, 1998, Computational Mechanics Publications, Boston, *Air Pollution*, **6**, 743–753, 1998b.

Jorquera, H., Palma, W. and Tapia, J.: Air quality at Santiago, An intervention analysis of air quality data at Santiago, Chile, *Atmospheric Environment*, **34**, 4073–4084, 2000.

Jorquera, H.: Air quality at Santiago, Chile: a box modelling approach I. Carbon monoxide, nitrogen oxides and sulfur dioxide, *Atmospheric Environment*, **36**, 315–330, 2002a.

Jorquera, H.: Air quality at Santiago, Chile: a box modelling approach II. $PM_{2.5}$, Coarse and PM_{10} particulate matter fractions, *Atmospheric Environment*, **36**, 331–344, 2002b.

Jorquera, H, Orrego, G. and Castro, J.: Trends in airr quality and population exposure in Santiago, Chile, 1989-2001, *International journal of Environment and Pollution*, **22**, 507-530, 2004.

Jorquera, H. and Rappenglück, B.: Receptor modelling of ambient VOC at Santiago, Chile, *Atmospheric Environment*, **38**, 4243-4263, 2004.

Kanaya, Y., Cao, R., Akimoto, H., Fukoda, M., Komazaki, Y., Yokouchi, Y., Koike, M., Tanimoto, H., Takegawa, N. and Konodo, Y.: Urban photochemistry in central Tokyo: 1. Observed and modelled OH and HO_2 radical concentrations during the winter and summer of 2004, *Journal of Geophysical Research*, **112**, D21312, doi:10.1029/2007JD008670, 2007.

Kanaya, Y., Pochanart, P., Liu, Y., Li, J., Tanimoto, H., Kato, S., Suthawaree, J., Inomata, S., Taketani, F., Okuzawa, K., Kawamura, K., Akimoto, H., and Wang, Z. F.: Rates and regimes of photochemical ozone production over Central East China in June 2006: a box model analysis using comprehensive measurements of ozone precursors, *Atmospheric Chemistry and Physics*, **9**, 7711-7723, 2009.

Kavouras, G., Lawrence, J., Koutrakis, P., Stephanou, E. G., Oyola, P.: Measurement of particulate aliphatic and polynuclear aromatic hydrocarbons in Santiago de Chile: source reconciliation, *Atmospheric Environment*, **33**, 4977-4986, 1999.

Kavouras, G., Koutrakis, P., Cereceda-Balic, F. and Oyola, P.: Source apportionment of PM10 and PM2.5 in five Chilean cities using factor analysis, *Journal of Air and Waste Management Association*, **51**, 451–464, 2001.

Kleffmann, J., Heland, J., Kurtenbach, R., Lörzer, J. C. and Wiesen, P.: A New Instrument (LOPAP) for the Detection of Nitrous Acid (HONO), *Environmental Science and Technology*, **9** (special issue 4), 48-54, 2002.

Kleffmann, J., Kurtenbach, R., Lörzer, J. C., Wiesen, P., Kalthoff, N., Vogel, B. and Vogel, H.: Measured and simulated vertical profiles of nitrous acid Part I: Field measurements, *Atmospheric Environment*, **37**, 2949-2955, 2003.

Kleffmann, J., Gavriloaiei, T., Hofzumahaus, A., Holland, F., Koppmann, R., Rupp, L., Schlosser, E., Siese, M. and Wahner, A.: Daytime Formation of Nitrous Acid: A Major Source of OH Radicals in a Forest, *Geophysical Research Letters*, **32**, L05818, doi: 10.1029/2005GL022524, 2005.

Kleffmann, J., Lörzer, J. C., Wiesen, P., Kern, C., Trick, S., Volkamer, R., Rodenas, M. and Wirtz, K.: Intercomparisons of the DOAS and LOPAP Techniques for the Detection of Nitrous Acid (HONO) in the Atmosphere, *Atmospheric Environment*, **40**, 3640-3652, 2006.

Kleffmann, J.: Daytime Sources of Nitrous Acid (HONO) in the Atmospheric Boundary Layer, *ChemPhysChem*, **8**, 1137-1144, 2007.

Kleffmann, J., and Wiesen, P.: Technical Note: Quantification of interferences of wet chemical HONO LOPAP measurements under simulated polar conditions, *Atmospheric Chemistry and Physics*, **8**, 6813–6822, 2008.

Kleinman L.: Ozone process insights from field experiments-part II: Observation-based analysis for ozone production, *Atmospheric Environment*, **34**, 2023-2033, 2000.

Kleinman, L. I., Daum, P. H.; Lee, Y. N., Nunnermacker, L. J., Springston, S. R., Weinstein-Lloyd, J. and Rudolph, J.: A comparative study of ozone production in five U.S metropolitan areas, *Journal of Geophysical Research*, **110**, D02301, doi: 10.1029/2004JD005096, 2005.

Koppmann, R., Johnen, F. J., Khedim, A., Rudolph, J., Wedel, A. and Wiards, B.: The influence of ozone on light nonmethane hydrocarbons during cryogenic preconcentration, *Journal of Geophysical Research*, **100**, 11383–11391, 1995.

Kurtenbach, R., Becker, K. H., Gomes, J. A. G., Kleffmann, J., Lörzer, J. C., Spittler, M., Wiesen, P., Ackermann, R., Geyer, A. and Platt, U.: Investigation of Emissions and Heterogeneous Formation of HONO in a Road Traffic Tunnel, *Atmospheric Environment*, **35**, 3385-3394, 2001.

Lei, W., de Foy, B., Zavala, M., Volkamer, R. and Molina, L. T.: Characterizing ozone production in the Mexico City Metropolitan Area: a case study using a chemical transport model, *Atmospheric Chemistry and Physics*, **7**, 1347–1366, 2007.

Leighton, P. A.: Photochemistry of Air Pollution, *Academic Press, New York*, 152–157, 1961.

Lelieveld, J., Dentener, F. J., Peters, W. and Krol, M. C.: On the role of hydroxyl radicals in the self-cleansing capacity of the troposphere, *Atmospheric Chemistry and Physics*, **4**, 2337-2344, 2004.

Levy, H.: II Normal atmosphere: Large radical and formaldehyde concentrations predicted. *Science*, **173**, 141-143, 1971.

Li, S., Matthews, J. and Sinha, A.: Atmospheric Hydroxyl Radical Production from Electronically Excited NO_2 and H_2O, *Science*, **319**, 1657-1660, 2008.

Liu, S. C., Cox, R. A., Crutzen, P. J., Ehhalt, D. H., Guicherit, R., Hofzumahaus, A., Kley, D., Penkett, S. A., Phillips, L. F., Poppe, D. and Rowland, F. S.: Group Report: Oxidizing Capacity of the atmosphere, in: The Changing Atmosphere, Rowland, F. S. and Isaksen, I. S. A., Eds., *Wiley, Chichester*, p. 219-232, 1988.

Lou, S., Holland, F., Rohrer, F., Lu, K., Bohn, B., Brauers, T., Chang, C. C., Fuchs, H., Häseler, R., Kita, K., Kondo, Y., Li, X., Shao, M., Zeng, L., Wahner, A., Zhang, Y., Wang, W., and Hofzumahaus, A.: Atmospheric OH reactivities in the Pearl River Delta –

China in summer 2006: measurement and model results, *Atmospheric Chemistry and Physics Discussion*, **9**, 17035-17072, 2009.

Madronich, S., and Garnier, C.: Impact of recent total ozone changes on tropospheric ozone photodissociation, hydroxyl radicals, and methane trends, *Geophysical Research Letters*, **19**, 465-467, 1992.

Mao, J., Ren, X., Brune, W. H., Olson, J. R., Crawford, J. H., Fried, A., Huey, L. G., Cohen, R. C., Heikes, B., Singh, H. B., Blake, D. R., Sachse, G. W., Diskin, G. S., Hall, S. R., and Shetter, R. E.: Airborne measurement of OH reactivity during INTEX-B, *Atmospheric Chemistry and Physics*, **9**, 163-173, 2009.

Maria del Rosario Sienra, M.: Oxygenated polycyclic aromatic hydrocarbons in urban air particulate matter, *Atmospheric Environment*, **40**, 2374-2384, 2006

Mihelcic, D., Holland, F., Hofzumahaus, A., Hoppe, L., Konrad, S., Müsgen, P., Pätz, H.-W., Schäfer, H.-J., Schmitz, T., Volz-Thomas, A., Bächmann, K., Schlomski, S., Platt, U., Geyer, A., Alicke, B. and Moortgat, G.: Peroxy radicals during BERLIOZ at Pabstthum: Measurements, radical budgets and ozone production, *Journal of Geophysical Research*, **108** (D4), 8254, doi: 10.1029/2001JD001014, 2003.

Milford, J., Gao, D., Sillman, S., Blossey, P., Russell, A.G.: Total reactive nitrogen (NO_y) as an indicator for the sensitivity of ozone to NO_x and hydrocarbons, *Journal of Geophysical Research*, **99**, 3533-3542, 1994.

Monod, A., Sive, B. C., Avino, P., Chen, T., Black, D. R., Rowland, F. S.: Monoaromatic compounds in ambient air of various cities: a focus on correlations between the xylenes and ethylbenzene, *Atmospheric Environment*, **35**, 135-149, 2001.

Nash, T.: The colorimetric estimation of formaldehyde by means of the Hantzsch reaction, *Biochemical Journal*, **55**, 416–421, 1953.

National Research Council (NRC), Committee on tropospheric Ozone Formation and Measurement: Rethinking the Ozone Problem in Urban and Regional Air Pollution, *National Academy Press*, 1991.

Ndour, M., D'Anna, B., George, C., Ka, O., Balkanski, Y., Kleffmann, J., Stemmler, K., and Ammann, M.: Photoenhanced uptake of NO_2 on mineral dust: Laboratory experiments and model simulations, *Geophysical Research Letters*, **35**, L05812, doi:10.1029/2007-GL032006, 2008.

Neftel, A., Blatter, A., Hesterberg, R., Staffelbach, Th.: Measurements of Concentration Gradients of HNO_2 and HNO_3 over a Semi-Natural Ecosystem, *Atmospheric Environment*, **30**, 3017-3025, 1996.

Niedojadlo, A.: Impact of NMVOC Emissions from Traffic and Solvent Use on Urban Air in Wuppertal-An Experimental study, *PhD Thesis, University of Wuppertal*, 2005.

Oanh, N. T. K., Martel, M., Pongkiatkul, P., Berkowicz, R.: Determination of fleet hourly emission and on-road vehicle emission factor using integrated monitoring and modeling approach, *Atmospheric Research*, **89**, 223–232, doi:10.1016/j.atmosres.2008.02.005, 2008.

Paulson, S. E., Flagan, R. C. and Seinfeld, J. H.: Atmospheric Photooxidation of Isoprene. Part 1: The Reactions of Isoprene with Hydroxyl Radical and Ground State Atomic Oxygen. *International Journal of Chemical Kinetics*, **24**, 79-102, 1992.

Paulson, S. E. and Orlando, J. J.: The reactions of ozone with alkenes: An important source of HO_x in the boundary layer, *Geophysical Research Letters*, **23**, 3727, 1996.

Paulson, S. E., Chung, M. Y. and Hasson, A. S.: OH radical formation from the gas-phase reaction of ozone with terminal alkenes and the relationship between structure and mechanism, *Journal of Physical Chemistry A*, **103** (41), 8125-8138, doi: 10.1021/jp991995e, 1999.

Perner, D. and Platte, U.: Detection of Nitrous Acid in the Atmosphere by Differential Optical Absorption, *Journal of Geophysical Rersearch*, **6** (12), 917-920, 1979.

Petersen, H. and Petry, N.: Formaldehyde-Allgemeine Situation, Nachweismethode, Einsatz in der Textilindustrie, *Melliand Textilberichte*, **66**, 285-295, 1985.

Prinn, G R.: The cleansing capacity of the atmosphere, *Annual Review of Environment and Resources*, **28**, 29-57, 2003.

Rappenglück, B., Oyola, P., Olaeta, I. and Fabian, P.: The evaluation of photochemical smog in the metropolitan area of Santiago de Chile, *Journal of Applied Meteorology*, **39**, 275-290, 2000.

Rappenglück, B., Schmitz R., Bauerfeind M., Cereceda-Balic, von-Baer D., Jorquera H., Silva Y. and Oyola P.: An Urban Photochemistry study in Santiago de Chile, *Atmospheric Environment*, **39**, 2913-2931, 2005.

Reidmiller, D. R., Fiore, A. M., Jaffe, D. A., Bergmann, D., Cuvelier, C., Dentener, F. J., Duncan, B. N., Folberth, G., Gauss, M., Gong, S., Hess, P., Jonson, J. E., Keating, T., Lupu, A., Marmer, E., Park, R., Schultz, M. G., Shindell, D. T., Szopa, S., Vivanco, M. G., Wild, O., and Zuber, A.: The influence of foreign vs. North American emissions on surface ozone in the US, *Atmospheric Chemistry and Physics Discussion*, **9**, 7927-7969, 2009.

Ren, X., Harder, H., Martinez, M., Lesher, R. L., Oliger, A., Simpas, J. B., Brune, W. H., Schwab, J. J., Demerjian, K. L., He, Y., Zhou, X. and Gao, H.: OH and HO_2 Chemistry in the Urban Atmosphere of New York City, *Atmospheric Environment*, **37**, 3639-3651, 2003.

Ren, X., Brune, W. H., Mao, J., Mitchell, M. J., Lesher, R. L., Simpas, J. B., Metcalf, A. R., Schwab, J.J., Cai, C., Li, Y., Demerjian, K. L., Felton, H. D., Boynton, G., Adams, A., Perry, J., He, Y., Zhou, X. and Hou, J.: Behavior of OH and HO_2 in the Winter Atmosphere in New York City, *Atmospheric Environment*, **40**, Supplement 2, 252-263, 2006.

Richter, P., Grino, P., Ahumada, I., Giordano, A.: Total element concnetration and chemical fractionation in airborne particulate matter from Santiago, Chile, *Atmospheric Environmenmt*, **41**, 6729-6738, 2007.

Rickard, A. R., Johnson, D., McGill, C. D. and Marston, G.: OH yields in the gas-phase reactions of ozone with alkenes, *Journal of Physical Chemistry*, **A103**, 7656–7664, 1999.

Rickard, A. R., Salisbury, G., Monks, P. S., Lewis, A. C., Baugitte, S., Bandy B. J., Clemitshaw, K. C. and Penkett, S. A.: Comparison of measured ozone production efficiencies in the marine boundary layer at two European coastal sites under different pollution regimes, *Journal of Atmospheric Chemistry*, **43**, 107-134, 2002.

Rohrer, F. and Berresheim, H.: Strong correlation between levels of tropospheric hydroxyl radicals and solar ultraviolet radiation, *Nature*, **442**, 184-187, doi:10.1038/nature04924, 2006.

Rubio, M. A., Lissi, E., Villena, G.: Nitrite in rain and dew in Santiago city, Chile. Its possible impact on the early morning start of the photochemical smog, *Atmopsheric Environment*, **36**, 293-297, 2002.

Rubio, M. A., Oyola, P., Gramsch, E., Lissi, E., Pizzaro, J. and Villena, G.: Ozone and peroxyacetylnitrate in downtown Santiago, Chile, *Atmospheric Environment*, **38**, 4931-4939, 2004.

Rubio, M. A., Lissi, E., Villena, G., Caroca, V., Gramsch, E. and Ruiz A.: Estimation of hydroxyl and hydroperoxyl radicals concentrations in the urban atmosphere of Santiago, *Journal of Chilean Chemical Society*, **50**, 2, 375-379, 2005.

Rubio, M. A., Zamorano, N., Lissi, E., Rojas A., Gutierrez L. and von Bare D.: Volatile carbonyl compounds in downtown Santiago, Chile, *Chemosphere*, **62**, 1011-1020, 2006a.

Rubio, M. A., Jose Guerrero, M., Villena, G., Lissi, E.: Hydroperoxides in dew water in downtown Santiago, Chile. A comparison with gas-phase values, *Atmospheric Environment*, **40**, 6165-6172, 2006b.

Rubio, M. A., Gramsch, E., Lissi, E., Villena, G.: Seasonal dependence of peroxyacetylnitrate (PAN) concentrations in downtown Santiago, Chile, *Atmosphera*, **20**, 319-328, 2007.

Rubio, M. A., Lissi, E., Villena, G.: Factors determining the concentration of nitrite in dew from Santiago, Chile, *Atmospheric Environment*, **42**,.7651-7656, 2008.

Rubio, M., Lissi, E., Villena, G., Elshorbany, Y. F., Kleffmann. J., Kurtenbach, R., Wiesen, P.: Simultaneous measurements of formaldehyde and nitrous acid in dews and gas phase in the atmosphere of Santiago, Chile, *Atmospheric Environment*, **43**, 6106 - 6109, doi:10.1016/j.atmosenv.2009.09.017, 2009a.

Rubio, M. A., Fuenzalida, I., Salinas, E., Lissi, E., Kurtenbach, R., Wiesen, P.: Carbon monoxide and carbon dioxide concentrations in Santiago de Chile associated with traffic emissions, *Environmental Monitoring and Assessment*, online publication, doi:10.1007/s10661-009-0789-9, 2009b.

Rutllant, J. and Garreaud R.: Meteorological air pollution potential for Santiago, Chile: towards an objective episode forecasting, *Environmental Monitoring and Assessment*, **34**, 223–244, 1995.

Salisbury, G., Monks, P. S., Bauguitte, S., Brandy, B. J. and Penkett, S. A.: A seasonal comparison of the ozone photochemistry in clean and polluted air masses at Mace Head, Ireland, *Journal of Atmospheric Chemistry*, **41**, 163-187, 2002.

Saltzmann, B. E.: Kinetic Studies of Formation of Atmospheric Oxidants, *Industrial & Engineering Chemistry*, **50**, 677-682, doi: 10.1021/ie50580a042, 1958.

Sarwar, G., Roselle, S. J., Mathur, R., Appel, W., Dennis, R. L. and Vogel, B.: A comparison of CMAQ HONO predictions with observations from the northeast oxidant and particle study, *Atmospheric Environment*, **42**, 5760-5770, 2008.

Sarwar, G., Pinder, R. W., Appel, K. W., Mathur, R. and Carlton, A. G.: Examination of the impact of photoexcited NO_2 chemistry on regional air quality, *Atmospheric Environment*, In Press, doi:10.1016/j.atmosenv.2009.09.012, 2009.

Saunders, S. M., Jenkin, M. E., Derwent, R. G. and Pilling, M. J.: Protocol for the development of the Master Chemical Mechanism, MCM V3: tropospheric degradation of non-aromatic VOC, *Atmospheric Chemistry and Physics*, **3**, 161-180, 2003.

Sax, S. N., Koutrakis P., Ruiz Rudolph P. A., Cereceda-Balic F., Gramsch, E. and Oyola P.: Trends in the elemental composition of fine particulate matter in Santiago, Chile, from 1998 to 2003, *Journal of Air and Waste Management Association*, **57**, 845-855, 2007.

Schmitz, R.: Modelling of air pollution dispersion in Santiago de Chile, *Atmospheric Environment*, **39**, 2035-2047, 2005.

Schreifel, J.: Emissions Trading in Santiago, Chile: *A Review of the Emission Offset Program of Supreme Decree No 4*, 2008. http://www.epa.gov/airmarkets/international/chile/et_santiago.pdf.

Schrimpf, W., Müller, K. P., Johnen, F. J., Lienaerts, K. and Rudolph, J.: An optimized method for airporn peroxyacetyl nitrate (PAN) measurements, *Journal of Atmospheric Chemistry*, **22**, 303-317, 1995.

Seguel, R., Morales, R. G. E., Leiva, M. A.: Estimation of primary and secondary organic carbon formation in $PM_{2.5}$ aerosol of Santiago City, Chile, *Atmospheric Environment*, **43**, 2125-2131, 2009.

Seinfeld, J. H. and Pandis, S. N.: Atmospheric chemistry and physics: From air pollution to climate change, *A Wiley-Interscience publications*, 1998.

Sheehy, P. M., Volkamer, R., Molina, L. T. and Molina, M. J.: Oxidative capacity of the Mexico City atmosphere-Part 2: A RO$_x$ radical cycling perspective, *Atmospheric Chemistry and Physics Discussion*, **8**, 5359-5412, 2008.

Shirley, T. R., Brune, W. H., Ren X., Mao J., Lesher R., Cardenas B., Volkamer R., Molina L. T., Molina M. J., Lamb B., Velasco E., Jobson T. and Alexander M.: Atmospheric oxidation in the Mexico City Metropolitan Area (MCMA) during April 2003, *Atmospheric Chemistry and Physics Discussion*, 6, 2753–2765, 2006.

Sienra, M., Rosazza, N. G. and Préndez, M.: Polycyclic aromatic hydrocarbons and their molecular diagnostic ratios in urban atmospheric respirable particulate matter, *Atmospheric Research*, **75**, 267-281, 2005.

Sienra, M.: Oxygentaed olycyclic aromatic hydrocarbons in urban air particulate matter, *Atmospheric Environment*, 40, 2374-2384, 2006.

Sillman, S.: The use of NO$_y$, H$_2$O$_2$ and HNO$_3$ as indicator for ozone-NO$_x$-hydrocarbon sensitivity in urban locations. *Journal of Geophysical Research*, **100** (14), 175-188, 1995.

Sillman, S., He, D., Cardelino, C., Imhoff, R.E.: The use of photochemical indicators to evaluate ozone - NOx - hydrocarbon sensitivity: case studies from Atlanta, New York and Los Angeles, *Journal of Air Waste Management Association*, **47**, 1030-1040, 1997.

Sillman, S., He, D., Pippin, M., Daum, P.H., Lee, J.H., Kleinman, L., Weinstein-Lloyd, J.: Model correlations for ozone, reactive nitrogen and peroxides for Nashville in comparison with measurements: implications for O3–NOx-hydrocarbon chemistry, *Journal of Geophysical Research*, **103**, 22629-22644, 1998.

Sillman, S.: The relation between ozone, NO$_x$ and hydrocarbons in urban and polluted rural environments, *Atmospheric Environment*, **33**, 1821-1845, 1999.

Sommariva, R., Bloss, W. J., Brough, N., Carslaw, N., Flynn, M., Haggerstone, A. -L., Heard, D. E., Hopkins, J. R., Lee, J. D., Lewis, A. C., McFiggans, G., Monks, P. S., Penkett, S. A., Pilling, M. J., Plane, J. M. C., Read, K. A., Saiz-Lopez, A., Rickard, A. R. and Williams, P. I.: OH and HO$_2$ Chemistry during NAMPLEX: roles of oxygenates, halogen oxides and heterogeneous uptake, *Atmospheric Chemistry and Physics*, **6**, 1135-1153, 2006.

Sjodin, A., Persson, K., Andreasson, K., Arlander, B., Galle, B.: On-road emission factors derived from measurements in a traffic tunnel, *International Journal of Vehicle Design*, **20**, 147–158, 1998.

Spindler, G., Hesper, J., Brüggemann, E., Dubois, R., Müller, Th. and Herrmann, H.: Wet Annular Denuder Measurements of Nitrous Acid: Laboratory Study of the Artefact Reaction of NO_2 with S(IV) in Aqueous Solutions and Comparison with Field Measurements, *Atmospheric Environment*, **37**, 2643-2662, 2003.

Stemmler, K., Bugmann, S., Buchmann, B., Reinmann, S. and Staehelin, J.. Large decrease of VOC emissions of Switzerland's car fleet during the past decade: results from a highway tunnel study, *Atmospheric Environment*, **39**, 1009-1018, 2005.

Stemmler, K., Ammann, M., Dondors, C., Kleffmann, J. and George, C.: Photosensitized Reduction of Nitrogen Dioxide on Humic Acid as a Source of Nitrous Acid, *Nature*, **440**, 195-198, 2006.

Stemmler, K., Ndour, M., Elshorbany, Y., Kleffmann, J., D'Anna, B., George, C, Bohn, B. and Ammann, M.: Light induced conversion of nitrogen dioxide into nitrous acid on submicron humic acid aerosol, *Atmospheric Chemistry and Physics*, **7**, 4237-4248, 2007.

Stockwell, W. R., Middleton, P., Chang, J. S and Tang, X.: The second generation regional acid deposition model chemical mechanism for regional air quality mdelling, *Journal of Geophysical Research*, **95**, 16343-16367, 1990.

Stockwell, W. R., Kirchner, F., Kuhn, M. and Seefeld, S.: A new mechanism for regional atmospheric chemistry modelling, *Journal of Geophysical Research*, **102** (D22), 25,847–25,880, 1997.

Stutz, J., Alicke, B. and Neftel, A.: Nitrous Acid Formation in the Urban Atmosphere: Gradient Measurements of NO_2 and HONO over Grass in Milan, Italy, *Journal of Geophysical Research*, **107** (D22), 8192, doi:10.1029/2001JD000390, 2002.

Stutz, J., Alicke, B., Ackermann, R., Geyer, A., Wang, S. H., White, A. B., Williams, E. J., Spicer, C. W. and Fast, J. D.: Relative humidity dependence of HONO chemistry in urban areas, *Journal of Geophysical Research*, **109**, D03307, doi:10.1029/2003JD004135, 2004.

Tie, X., Madronich, S., Li, G., Ying, Z., Weinheimer, A., Apel, E., and Campos, T.: Simulation of Mexico City plumes during the MIRAGE-Mex field campaign using the WRF-Chem model, *Atmospheric Chemistry and Physics*, **9**, 4621-4638, 2009.

Thompson, A. M.: The oxidizing capacity of the earth's atmosphere: probable past and future changes, *Science*, **256**, 1157-1165, 1992.

Trainer, M., Parrish, D. D., Buhr, M. P., Norton, R. B., Fehsenfeld, F. C., Anlauf, K. G., Bottenheim, J. W., Tang, Y. Z., Weibe, H. A., Roberts, J. M., Tanner, R. L., Newman, L., Bowersox, V. C., Meagher, J. T., Olszyna, K. J., Rodgers, M. O., Wang, T., Berresheim, H., Demerjian, K. L. and Roychowdhury, U. K.: Correlation of ozone with NOy in photochemically aged air, *Journal of Geophysical Research*, **98** (D2), 2917-2925, 1993.

Trier, A. and Firingueti, L.: A time series investigation of visibility in an urban atmosphere—I, *Atmospheric Environment*, **28**, 991–996, 1994.

Vogel, B., Vogel, H., Kleffmann, J. and Kurtenbach, R.: Measured and Simulated Vertical Profiles of Nitrous Acid, Part II – Model Simulations and Indications for a Photolytic Source, *Atmospheric Environment*, **37**, 2957-2966, 2003.

Volkamer, R., Sheehy, P. M., Molina, L. T. and Molina, M. J.: Oxidative capacity of the Mexico City atmosphere – Part 1: A radical source perspective, *Atmospheric Chemistry and Physics Discussion*, **7**, 5365–5412, 2007.

Volz-Thomas, A., Xueref, I., Schmitt, R.: An Automatic gas Chromatograph and calibration system for ambient measurements of PAN and PPN, *Environmental Science and Pollution Research*, **4**, 72–82, 2002.

Wang, Y., Hao, J., McElroy, M. B., Munger, J. W., Ma, H., Chen, D., and Nielsen, C. P.: Ozone air quality during the 2008 Beijing Olympics: effectiveness of emission restrictions, *Atmospheric Chemistry and Physics*, **9**, 5237-5251, 2009.

Wayne, R. P. (ed.): The nitrate radical: Physics, chemistry, and the atmosphere, *Atmospheric Environment*, **25**, 1-203, 1991.

Wennberg, P. O. and Dabdub, D.: Rethinking ozone production, *Science*, **319**, 1624-1625. doi:10.1126/science.1155747, 2008.

West, J. J., Naik, V., Horowitz, L. W., and Fiore, A. M.: Effect of regional precursor emission controls on long-range ozone transport – Part 1: Short-term changes in ozone air quality, *Atmospheric Chemistry and Physics*, **9**, 6077-6093, 2009.

WHO Air Quality Guidelines: Global Update 2005. Particulate matter, ozone, nitrogen dioxide and sulfur dioxide. *WHO Regional Office for Europe*, Scherfigsvej, 8 DK-2100 Copenhagen Ø, Denmark, 2005.

Woolfenden, E. A. and McClenny, W. A.: Determination of Volatile organic compounds in the ambient air using active sampling onto sorbent tubes. Compendium method TO-17. Compendium of methods for the determination of toxic organic compounds in ambient air, second edition. *Office of research and development, US environmental protection agency*, Cincinnati, OH 45268, 1999.

Yoshino, A., Sadanagab, A., Watanabea, K., Katoa, S., Miyakawaa, Y., Matsumotoc, J. and Kajiia, Y.: Measurement of total OH reactivity by laser-induced pump and probe technique-comprehensive observations in the urban atmosphere of Tokyo, *Atmospheric Environment*, **40**, 7869-7881, 2006.

Zawala, M., Lei, W., Molina, M., Molina, L. T.: Modeled and observed ozone sensitivity to mobile source emissions in Mexico City, *Atmospheric Chemistry and Physics*, **9**, 39-55, 2009.

Zhang, J., Wang, T., Chameides, W. L., Cardelino, C., Kwok, J., Blake, D. R., Ding, A. and SO, K. L.: Ozone production and hydrocarbon reactivity in Hong Kong, Southern China, *Atmospheric Chemistry and Physics*, **7**, 557-573, 2007.

Zhang, N., Zhou, X., Shepson, P. B., Gao, H., Alaghmand, M., and Stirm, B.: Aircraft measurement of HONO vertical profiles over a forested region, *Geophysical Research Letters*, **36**, L15820, doi:10.1029/2009GL038999, 2009.

Zhou, X., Civerolo, K., Dai, H., Huang, G., Schwab, J. and Demerjian, K.: Summertime Nitrous Acid Chemistry in the Atmospheric Boundary Layer at a Rural Site in New York State, *Journal of Geophysical Research*, **107** (D21), 4590, doi:10.1029/2001JD001539, 2002.

Zhou, X., Gao, H., He, Y., Huang, G., Bertman, S. B., Civerolo, K and Schwab, J.: Nitric Acid Photolysis on Surfaces in Low-NO_x Environments: Significant Atmospheric

Implications, *Geophysical Research Letters*, **30** (23), 2217, doi: 10.1029/2003GL018620, 2003.

Die VDM Verlagsservicegesellschaft sucht für wissenschaftliche Verlage abgeschlossene und herausragende

Dissertationen, Habilitationen, Diplomarbeiten, Master Theses, Magisterarbeiten usw.

für die kostenlose Publikation als Fachbuch.

Sie verfügen über eine Arbeit, die hohen inhaltlichen und formalen Ansprüchen genügt, und haben Interesse an einer honorarvergüteten Publikation?

Dann senden Sie bitte erste Informationen über sich und Ihre Arbeit per Email an *info@vdm-vsg.de*.

Sie erhalten kurzfristig unser Feedback!

VDM Verlagsservicegesellschaft mbH
Dudweiler Landstr. 99   Telefon +49 681 3720 174
D - 66123 Saarbrücken   Fax   +49 681 3720 1749
www.vdm-vsg.de

Die VDM Verlagsservicegesellschaft mbH vertritt

Printed by Books on Demand GmbH, Norderstedt / Germany